SAVING OUR SEEDS

Photo by Small House

SAVING OUR SEEDS
··· The Practice & Philosophy ···

Bevin Cohen

Copyright © 2019 Bevin Cohen
Sanford, MI
www.smallhousefarm.com

Cover and Page Design by **Heather Cohen**
Photographs courtesy of **Baker Creek Heirloom Seed Company** unless otherwise noted

Cover Photo Credit
TOP *Baker Creek Heirloom Seed Company*
BOTTOM *Kari Witthuhn-Henning from Appleton Seed Library*

All rights reserved.
No part of this book may be reproduced in any form or by any means without the prior written permission of the publisher, except brief quotations in connection with reviews written specifically for inclusion in magazines or newspapers, or limited excerpts strictly for personal use.

ISBN-13: 978-0-578-55589-8

To my beautiful wife Heather.
Without her, nothing would be possible.
She is truly the most patient woman on the planet.

Table of Contents

Adzuki 14
Bean, Common 16
Beet 22
Broccoli 24
Brussels Sprouts 26
Cabbage 28
Carrot 34
Cauliflower 36
Chard 38
Chickpea 44
Collard 46
Corn 48
Cowpea 54
Cucumber 60
Eggplant 62
Fava Bean 68
Ground Cherry 70
Kale 76
Kale, Russian 78
Kohlrabi 80
Lentil 82

Lettuce 84
Lima Bean 90
Melon 92
Mustard Green 94
Okra 96
Parsnip 102
Peanut 104
Pea 108
Pepper 110
Potato 112
Radish 118
Runner Bean 120
Soy Bean 122
Spinach 128
Squash 130
Sunflower 140
Tepary Bean 142
Tomatillo 144
Tomato 146
Turnip 152
Watermelon 154
Wheat 166

Table of Contents

The Practice

- Flower Structure and Pollination 170
- Hand Pollination 172
- Dry Seed vs Wet Seed 176
- Threshing and Winnowing 182
- Overwintering Biennials 184
- Labeling and Storage 186

The Philosophy

- Seeds as Teachers 13
- A Handful of Stories 30
- Hello Dahlia 40
- Historic Gardens, Frozen in Time 72
- How Diversity Fills Our Plate 86
- Standing Amongst the Corn 106
- Potential in an Empty Field 124
- Modern Trade Routes 166

The Seedkeeper's Tales

- Russell Crow 18
- Rafael Mier 50
- Angie Lavezzo 56
- Mehmet Öztan 64
- Chris Smith 98
- Curzio Caravati 114
- Sarah Tomac 134
- Laura Flacks-Narrol 148
- Rob Mcelwee 156

Photo by Baker Creek Heirloom Seed Company

Foreword

Seed saving is more than a hobby or a pastime. It is a way of life that lends itself to preserving the past, as well as preparing for the future. Bevin Cohen is an avid gardener and seed saver whose ambitious efforts to create seed libraries across the Midwest and beyond are helping others to preserve the past and prepare for the future.

My own adventure as a seed grower and seed saver began when I was a very small child in a family that valued gardening. I planted my first seeds when I was only 3 years old. Being home schooled, I learned to read by looking at seed catalogs and gardening magazines. When I was 12 years old, my family moved from the Idaho Valley to Missouri with its longer growing season and more suitable growing climate. By that time, I was already well into my hobby of saving seeds. I already knew that I somehow wanted to work with seeds later in life, but I had no idea if that dream would become a reality or where it would take me.

I realized at an early age that many of the seed varieties I liked in one seed catalog could not always be found in the next year's edition. Over the years, I learned that varieties were "disappearing" and were no longer available. This realization led to my interest in preserving seed diversity. That interest in saving seed diversity led to my founding Baker Creek Heirloom Seed Company, which specializes in finding rare seeds from all over the world and making them available to gardeners everywhere.

I have witnessed the growth in the garden movement that includes people of all ages from small children up through senior adults. Many of those gardeners are new to the gardening scene and planting their first gardens. Many others are "returnees" who once gardened, gave it up for various reasons, and have returned to experience the benefits of growing their own food. Still others are older folks who have never grown a garden before but now have an interest in producing their own food, which they perceive to be a healthy alternative to supermarket produce.

Foreword

Part of establishing a growing seed company has been an interest in learning about seeds and the stories behind those seeds and the people growing them. Early in my career, I began bringing together people who shared those interests. We first began holding "festivals" on the Baker Creek farm in the year 2000 and moved on to eventually helping create the National Heirloom Exposition in California in 2011. Key to all of those are the speakers who come and share their knowledge of gardening and seed saving, which is where Bevin enters the picture. I first met Bevin Cohen when he spoke at our 2018 National Heirloom Expo in Santa Rosa, California, and then later at our 2019 Spring Planting Festival that draws 10,000 people to our farm near Mansfield, MO. I realized that his ideals and values about seed saving are similar to mine. His message to his physical audience is the same as the message to his audience of readers: SAVING OUR SEEDS is important. Both he and I have had the opportunity to meet some knowledgeable and influential advocates of seed preservation.

Bevin has a natural ability to get people interested in gardening with heirloom seeds, saving seeds, and sharing their seed stories. Being an active advocate of community seed sharing, he travels across the country with his wife and two children educating people about the importance of genetic diversity. Because he and his family work to maintain the diversity of heirloom crops on their own farm, he successfully shares information about pollination basics and seed saving techniques. Similar to his natural capability to interest people when talking directly to them, he also has a comparable ability to draw their interest to the teachings in his book, SAVING OUR SEEDS, The Practice & Philosophy. Just as he and his wife live a simple life on their Small House Farm, Bevin presents his ideas simply and concisely in this book.

Foreword

I have appreciated getting to know Bevin Cohen, as we have shared seed saving ideas and stories about seeds and their growers. He has a genuine interest in learning the stories about various seeds and sharing those stories with others. Just as I enjoy bringing together gardeners and seed savers to discuss ideas about preserving the seeds of the past to prepare for the future, Bevin Cohen is a master of doing that same thing in this book. I consider him to be a valued colleague in the world of gardening with heirloom seeds.

Jere Gettle
Founder and owner of Baker Creek Heirloom Seed Company

Photo by Rareseeds.com

Photo by Baker Creek Heirloom Seed Company

Seeds as Teachers
By Bevin Cohen

• •

As gardeners, we all know that our annual adventures in the soil can leave us tired, dirty and sometimes frustrated… but we also know the rush that comes with seeing the first sprout pushing its way free from the soil, the sheer pleasure of beautiful flowers waving in the wind on a warm summer day and the indescribable ecstasy of a bountiful harvest; baskets full of produce covering our dinner tables and filling our pantries.

We also know that while the garden may challenge and reward us, if we take the time to observe and listen, it can also be our greatest teacher. Our gardens will teach us patience and humility. We learn about sharing and reciprocity. We also learn about the significance and importance of death as part of the cycle of life. When we harvest and save seeds from our garden crops, we become an active part of this cyclical nature of being. As gardeners and seed savers, we are stewards of the soil, keepers of the seed, singers of the song of life; no other work on this Earth is as important.

As you enter the realm of the seed saver, this book will serve as your guide. Within its pages you will find instructions on how and when to harvest seed from dozens of different species as well as how to process and store these miraculous bundles of life. We will delve into the philosophy of seed keeping and its role in the resurgence of localized food systems. You will read the stories of many gifted seed stewards and learn about what inspired each of them to take their first steps along this path.

As you turn these pages, you join us in the circle and together we journey forward…

ADZUKI BEAN

Vigna angularis

Family: Fabaceae

Pollination:
Self-pollinating though occasionally pollinated by insects

Isolation:
10-20 feet

DRY SEED

A **ANNUAL**

Photo by Baker Creek Heirloom Seed Company

ADZUKI BEAN

The Adzuki bean is little known in North America, although it has a very long and noble history of cultivation and use in Asia. It is the most common bean used in preparing the famous bean paste of the orient, which is sweetened and used as an ingredient in deserts. Adzuki beans are the preferred beans for bean paste because of their fine consistency

Adzuki beans aren't found in nearly as many gardens as some of its relatives, but I think this is only due to a lack of familiarity. These beans are easy to grow and are simply delicious! Adzuki beans are of the same genus as a number of garden crops including the well-known cowpea, and its seeds are harvested and processed in a similar way. It's believed that *Vigna angularis* was domesticated over 7,500 years ago in South America and eventually made its way to Asia and India in the sixteenth century thanks to the Spaniards. Many cultivars boast a beautiful and uniform dark red color but there are varieties available on the market with seeds that are white, black and gray as well as mottled patterns.

These beans are a longer season crop and will need about 120 days from planting for harvest of the seeds for dry use, or next year's garden seed. Allow the pods to mature on the plant and dry down until they are crisp and brown. At this point, they can be harvested, threshed and winnowed using the same techniques used to process other dry bean crops. More information on this can be found on page 182.

SAVING OUR SEEDS: The Practice & Philosophy

BEAN COMMON

Phaseolus vulgaris

Family: Fabaceae

Pollination:
Self-pollinating though occasionally pollinated by insects depending upon variety

Isolation:
10-20 feet

DRY SEED

A ANNUAL

Photo by Baker Creek Heirloom Seed Company

Blue Lake Bush Bean

Originally developed in the early 1900's for canning. They get their name from the Blue Lake area near Ukiah, California, where they originated. By the mid-to-late 1920's, Blue Lake Bush had been developed into a string-less variety for use as a green bean.

One of the most popular of garden crops, the common bean is also one of the simplest species to save seeds from. Cross pollination between most varieties is quite rare and the harvesting and processing of the dried beans is quite straightforward. While I enjoy the ease at which I can grow and harvest beans, the true thrill of including *Phaseolus vulgaris* in the gardens is the absolute beauty and diversity found in its seeds! Minnesota plant breeder Robert Lobitz referred to the common bean as "a poor man's jewels" and he couldn't have been more right. The astounding array of colors and shapes found throughout this species is anything but common in my eyes, each seed is truly a work of art. An entire volume could be dedicated to the topic of beans; their history, beauty and many culinary uses.

The growth habit of the common bean also varies from one variety to the next; some are determinate bush types, some are indeterminate (pole beans) and others are semi-determinate half-runner types. It is the half-runner varieties that are most commonly utilized in a three sisters style planting as fully indeterminate beans can overcome and even pull down the cornstalks they are using for support!

Bean seeds are harvested and processed as dry seeds after the plants have died back and the pods are brown and crisp. More information on the threshing and winnowing of dry seeds can be found on page 182. If there is threat of rain or frost before your bean seeds are ready to harvest, pods that have begun to yellow can be picked individually and left on a screen to dry or entire plants can be pulled and hung in a cool dark area with good airflow to continue drying down.

SAVING OUR SEEDS: The Practice & Philosophy

SEEDKEEPER'S TALE

RUSSELL CROW
ILLINOIS, USA
FOUNDER, THE LITTLE EASY BEAN NETWORK

When I stop to think about what led me to the days that I began growing and collecting beans in a big way, I have to go all the way back to my childhood. I believe that every little incremental step on your path, paired with the correct environment at the right moment in time, leads you down those roads of your most major life experiences.

I grew up in what is now one of the Chicago suburbs, Lisle, Illinois. In the early 1950's it was very much a rural town. There I remember my fathers gardens, his Kentucky Wonder pole beans and many other things he grew. One day I said, "Dad I want a garden of my own". So he prepared a small 18 by 24 inch piece of ground that I planted completely in radishes. That was my first garden, and apparently gardening appealed to me. The journey down that gardening road had begun. Gardening has not been a constant activity for me, but something that would draw me back like a magnet from time to time.

My dad passed away in 1956 when I was 10 and in 1958 my mother, with some discussion, allowed my brother-in-law to build us a new house on the southeast side of our town. No longer would we have to stoke the fire of a huge coal burning stove in the basement. It was now modern, automatic fuel oil heat in the winter. A new home of lower maintenance.

SAVING OUR SEEDS: The Practice and Philosophy

In the spring of 1959 I asked my mom if I could have a garden near the edge of our backyard. "Yes that would be nice," she said. So I prepared by hand a 15' by 70' garden space. We grew the usual garden vegetables: tomatoes, onions, peppers, snap beans, sweet corn, cucumbers, radishes, and carrots. My very own first garden would last until about the mid 1960's.

In those days, I also had gotten into going door to door selling seeds in the spring from the American Seed Company. One spring I had some zinnia and Topcrop snap bean seed left over after the peak selling time. I planted the zinnias in a flowerbed near the dining room, which faced south. One day, in late summer, I noticed one of my beautiful zinnia blossoms had turned brown and dry. So I snipped it off and decided to take the blossom apart. To my amazement I found those same arrowhead shaped seeds at the ends of petals that were in my American Seed packet in the spring.

As for the good amount of Topcrop snap bean seed left over, I didn't know anything about succession planting, so I planted all of the snap bean seed at once; 8 rows, 15 feet long. I had no idea of the snap bean harvest that was to come. My mother showed me how to prepare them for freezing and we were eating plenty of green beans with our summer meals. Needless to say, with time, the job of picking and preparing those beans for the freezer wore on me. So I stopped picking them after awhile. Then one day in late summer while mowing the lawn near the garden, I noticed some brownish things hanging from the bean plants. So I stepped into the bean patch and picked one to see what it was; a bean pod light tan and dry to the touch. I squeezed it and with a slight cracking sound, it opened a bit. So I opened it the rest of the way and there were the same Topcrop seeds that I planted from those seed packets in the spring.

In 1974 I got married and wanted to move out of the Chicago suburban environment, so with money I had saved working a second job, I purchased six acres in rural Boone county in Capron, Illinois, east of Rockford. There I had a new small 900 square foot ranch style house built. I worked at a die-casting plant in Woodstock, Illinois and my checks were not large. I was moved frequently from job to job and could never get proficient at any one of them. So with six acres it was time to start gardening to supplement the grocery list. I had remembered my mother preparing pinto beans for my dad in the 1950's as he was a diabetic. Instead of the pasta and potatoes the rest of the family ate at times, he would eat beans. They digest slowly and are low glycemic, impacting your blood sugar very gradually. I remembered how well they stored and I also remembered how I could grow them myself. So with an array of seed catalogs, I purchased a number of snap bean varieties and dry bean varieties. I did notice in passing the variation in seed coat coloring and designs some of the beans had; pintos, horticultural beans, great northerns and various mottled seeds of the snap beans.

Living in our rural setting, my wife and I would do our grocery shopping in one of the nearby towns. Harvard, Illinois had a nice little drug store with a beautiful magazine rack that I took notice of. I found Organic Gardening and Farming magazine there and purchased each new issue every month.

In the January 1978 issue of Organic Gardening and Farming, there was an article about a man from Lynnfield, Massachusetts by the name of John Withee. The article was about his bean collection and the network of gardeners (Wanigan Associates) he had developed to help him keep his collection alive. I found the article quite compelling and with the information presented in the article, I sent off my $5.00 for a copy of his bean catalog, which I still have to this very day. I studied his catalog for about two weeks and was fascinated by the descriptions of the beans in his catalog. Often he would describe seed colors and patterns of various varieties. So with a good degree of excitement I sent in some more dollars and a list of 35 varieties found in his little bean book.

Upon the arrival of John Withee's parcel in my mailbox several weeks later, I became smitten with the contents. How easy to grow, useful, beautiful, and varied were all those incredible beans. It was like looking at some exotic deep space photo from the Hubble telescope. I just knew I was going to acquire more of the them. There was no way out, I was hooked. For several years I grew his beans as I acquired more from him. I sent John back fresh seed each autumn to do my little part in helping him replenish his seed holdings.

Along with the early purchases of some 15 or so commercial varieties, John Withee's heirloom beans became the basis of some of my own original named varieties. Many of them still exist and are grown by other gardeners today; named beans like Blue Jay, Pawnee, Candy, and Kishwaukee Yellow. I believe in preserving heirloom bean varieties, but I am not a purist in that endeavor. I also believe in allowing promising and useful new varieties to flourish alongside the tried and true.

Later, in 1978, I also discovered an advertisement in the magazine Mother Earth News for a seed network called Seed Savers Exchange (SSE). In the 1979 Seed Savers yearbook, I listed those same 35 beans I first acquired from John Withee's Wanigan Associates. SSE is a preservation organization located on an 880 acre farm in Decorah, Iowa. They are dedicated to distributing seeds through their network and keeping heirloom vegetable varieties alive. I am still a member of SSE.

I developed my own bean website in 2012, which revolves around my collection, called A Bean Collectors Window. I also have my own small network of bean growers called "The Little Easy Bean Network".

This is the gardening road I have traveled along, doing just a small part in saving old but never obsolete bean varieties. It takes many of us for the task. What better reasons could one think of to be a seed saver? Humans live by variety and not by the drudgery of a monotone existence. It is the same with all our cultivars; a useful resource to nourish the body and breathtaking beauty to feed the spirit. What a boring and colorless world this becomes when any of it disappears for the lack of attention to preserve it. We are all part of the same creation and this resource is just as deserving of its life as we are, for as long as the earth is still here. Shame on any generation that would allow it to pass away. What has been left to us is truly a treasure and a gift from the past.

BEETS

Beta vulgaris

Family: Amaranthaceae

Pollination: Wind Pollinated

Isolation: 800 feet – 1 mile

DRY SEED

B BIENNIAL

OVER-WINTERING REQUIRED ...SEE PG 184

Photo by Baker Creek Heirloom Seed Company

Cheltenham Green Top Beet

Bred by A H Cook of Cheltenham, England. A heavy cropping tap-rooted old variety that has been around since before the 1880's. This beet has long tapering roots and a very well deserved reputation for outstanding flavor, texture and eating qualities.

As a child, I never could seem to enjoy the flavor of beets when they were offered at dinner time, or any time for that matter. I don't know if it was the texture or the presentation, but as an adult I lament over the time I wasted as a youth… time I could have spent eating more beets!

The species *Beta vulgaris* includes the familiar table beets and fodder beets, as well as sugar beets and even Swiss chard. As this species is wind pollinated, great care must be taken to keep your variety isolated to avoid crossing. The pollen produced by these plants is very small and light, so the easiest way to ensure purity is to only allow one variety to flower each year. Beets are biennial and need to be over wintered as described on page 184.

With some advanced planning and an alternating schedule, you could maintain a few different varieties of your favorite beets and chards. It's interesting to note that the leafy phenotype of *Beta vulgaris* was likely domesticated 2,000 years ago, the root vegetable we call beets wasn't developed until around the 16th century. The well known sugar beet wasn't bred until about 200 years ago and has since become a commercial commodity that many of us are familiar with.

You will want to plan extra space for your second year beetroot plants as they can grow quite large and will likely also require staking. When the fruits that form on the branches begin to turn brown and dry, they can be collected and cleaned like other dry seeds. You can learn more about processing dry seeds on page 182.

SAVING OUR SEEDS: The Practice & Philosophy

BROCCOLI

Brassica oleracea

Family: Brassicaceae

Pollination: Insect Pollinated

Isolation: ½ mile

DRY SEED

A ANNUAL

Photo by Baker Creek Heirloom Seed Company

Early Purple Sprouting Broccoli

An English heirloom variety, bred for overwintering. It was once quite popular and grown extensively throughout Europe. According to MM. Vilmorin-Andrieux in "The Vegetable Garden" (1885), there were more than forty different forms grown in England alone.

Broccoli is one member of a plant species that is possibly one of the most diverse food crops in cultivation. Other members of this species include cabbage, kale, collards, Brussels sprouts, kohlrabi and cauliflower. *Brassica oleracea* is typically considered a biennial but some of the plants in this species, including broccoli, are annuals and will produce seed in the first season they are grown.

Broccoli was developed over time as farmer selected from their cole crops for plants that produced edible flower buds and, if left to mature, the heads of broccoli will develop into flowers stalks and eventually seed pods. While the plants of this species are all self-incompatible, they will readily cross pollinate amongst themselves and caution must be taken to avoid this. Since many of these plants are biennial, the grower can enjoy a diverse selection in one season while allowing one variety from the previous year to go to seed. Although the many crops of *B. oleracea* may have various growing needs, they essentially all produce seed in the same manner and can be processed with similar techniques.

Seeds of *Brassica oleracea* plants are produced in small pods call siliques. When the seeds are mature, these siliques will dry and turn a light brown color. Care must be taken when gathering the dried pods as they can easily shatter and the seeds will be lost. Once collected, the pods can easily be flailed or threshed to remove the seeds which can then be easily winnowed or run through screens to remove the debris. More details about threshing and winnowing dry seeds can be found on page 182.

BRUSSELS SPROUTS

Brassica oleracea

Family: Brassicaceae

Pollination: Insect Pollinated

Isolation: ½ mile

DRY SEED **B BIENNIAL**

OVER-WINTERING REQUIRED ...SEE PG 184

Photo by Baker Creek Heirloom Seed Company

BRUSSELS SPROUTS

Brussels sprouts are one member of a plant species that is possibly one of the most diverse food crops in cultivation. Other members of this species include broccoli, kale, collards, cabbage, kohlrabi and cauliflower. *Brassica oleracea* is typically considered a biennial but some of the plants in this species are annuals and will produce seed in the first season they are grown. Brussels sprouts are a biennial and will require overwintering in order to flower and produce seed. More information on this process can be found on page 184.

Brussels sprouts were likely developed from cabbage and produce tightly cupped leaves along a prominent central stem. It's commonly believed that Brussels sprouts were developed in the Belgian town of the same name, sometime in the sixteenth century. The sprouts are said to be much sweeter after being exposed to a light frost.

While the plants of this species are all self-incompatible, they will readily cross pollinate amongst themselves and caution must be taken to avoid this. Since many of these plants are biennial, the grower can enjoy a diverse selection in one season while allowing one variety from the previous year to go to seed. Although the many crops of *B. oleracea* may have various growing needs, they essentially all produce seed in the same manner and can be processed with similar techniques.

Seeds of *Brassica oleracea* plants are produced in small pods call siliques. When the seeds are mature, these siliques will dry and turn a light brown color. Care must be taken when gathering the dried pods as they can easily shatter and the seeds will be lost. Once collected, the pods can easily be flailed or threshed to remove the seeds which can then be easily winnowed or run through screens to remove the debris. More details about threshing and winnowing dry seeds can be found on page 182.

CABBAGE

Cabbage is one member of a plant species that is possibly one of the most diverse food crops in cultivation. Other members of this species include broccoli, kale, collards, Brussels sprouts, kohlrabi and cauliflower. *Brassica oleracea* is typically considered a biennial but some of the plants in this species are annuals and will produce seed in the first season they are grown. Cabbage is a biennial and will require overwintering in order to flower and produce seed. More information on this process can be found on page 184.

The head of a cabbage plant is formed by tightly cupped leaves surrounding the terminal bud. After the plants are overwintered, these leaves will open and the flower stalk will then emerge. There is quite a bit of variation between cabbage varieties; some are white and green, others are red or purple. Some cabbages have smooth leaves while some have a more savoyed texture. There are also varieties of Chinese cabbage, such as Napa, but these are actually *Brassica rapa* and cross pollination with these types is of no concern.

While the plants of this species are all self-incompatible, they will readily cross pollinate amongst themselves and caution must be taken to avoid this. Since many of these plants are biennial, the grower can enjoy a diverse selection in one season while allowing one variety from the previous year to go to seed. Although the many crops of *B. oleracea* may have various growing needs, they essentially all produce seed in the same manner and can be processed with similar techniques.

Seeds of *Brassica oleracea* plants are produced in small pods call siliques. When the seeds are mature, these siliques will dry and turn a light brown color. Care must be taken when gathering the dried pods as they can easily shatter and the seeds will be lost. Once collected, the pods can easily be flailed or threshed to remove the seeds which can then be easily winnowed or run through screens to remove the debris. More details about threshing and winnowing dry seeds can be found on page 182.

SAVING OUR SEEDS: The Practice & Philosophy

A Handful of Stories
by Bevin Cohen

• •

A handful of seed is a handful of stories. These stories carry our history and cultures and, ultimately, our survival within them. Even the tactile sensation of seeds in one's palm can trigger the memories and emotions associated with that particular variety, as if transporting one back in time. Truly every seed is a time capsule.

Through my many years of seed keeping work, I have had the great honor of hearing a number of these special tales; some tracing a seed's journey back hundreds of years while others are sentimental, closely following the history of a family or neighborhood. While seed stories come in many forms, all are significant and deserving of preservation. Throughout time, many of these stories have been passed down through oral tradition, shared over meals or in the fields from one seed keeper to the next. In more modern times, many of us 'heirloom activists' have taken on the task of preserving this history through the written word and even audio recordings. These stories must be saved and passed down like the seeds themselves; there is no difference between the two.

Last fall I was in southern Indiana to lead a seed harvesting workshop at a community seed library in Jefferson County. While I was there I met some great people and had the opportunity to swap seeds with many of them. After the event I got to talking with a lady named Anne-Marie who lived in the next town over, just a ten minute drive along the Ohio River. During our conversation she briefly mentioned a pepper variety that she has been growing in her gardens that she originally brought back from Brazil in 2007. I excitedly responded that I'd be interested in trying to grow some of her peppers as well and I gave her my contact information so we could stay in touch. It would be three months before I would hear from Anne-Marie again. On January 14th, I received an email letting me know that she had finally finished processing the pepper seeds from her fall harvest and my package was ready to go out in the mail! The special seeds that Anne-Marie was sending my way were for a variety known as Pimento Coumari. I was certainly excited that my new seed keeping friend was sending me a sample of her special seeds but what was especially thrilling for me was the story that came along with the precious package.

As I mentioned earlier, Anne-Marie has been growing Pimento Coumari at her home in Indiana since she brought the seeds back with her after a trip to Brazil in 2007. She received the seeds from a work colleague's father, Mr. Pires, who had been growing these peppers on his balcony in the Brazilian city of Americana for many years. She also shared with me how these very hot peppers have a long history of use, even in the most remote areas of the country, and are used to liven up traditional bean and rice dishes. The connection between the peppers themselves and the town in which Anne-Marie acquired them from Mr. Pires is rather insignificant, but this was my first time learning about the city of Americana, Brazil.

The town was originally populated in 1866 by Confederate Americans that were unhappy with the outcome of the American Civil War. It was originally known as Village of the Americans until 1904. Again, the connection between this city and the pepper variety grown by Mr. Pires, and later Anne-Marie, is truly only coincidence, but the Village of the Americans was certainly new to me and this was a great opportunity for an unexpected history lesson!

As of yet, I have been unable to find any other information about, or references to, this pepper variety. I was able to find a listing for 'Cumari do Para' peppers Online, but I do not believe it is possible that these are the same fruits. Anne-Marie's pepper is described to me as "small, with purple

on the stems and leaf veins, peppers are changing color purple to red when ripe" and the online listing for 'Cumari do Para' says that this variety boasts super-hot, tiny pea-sized yellow peppers. Interestingly enough, both of these varieties are originally from Brazil. I plan to grow out the seeds gifted to me in an upcoming season and the hunt for more information about Anne-Marie's Pimento Coumari will continue!

Another brief but interesting seed story that I'd like to share also comes from Indiana, although this time from the northern part of the state. I had been invited to the small town of Goshen to speak to the community in hopes of inspiring folks to organize a seed library program. To aid in the event's appeal to the community, we also hosted a small seed swap during my visit. These types of programs always tend to draw some interesting people and this particular event was no different. One of the many fine folks I met in Goshen that day was a lady by the name of Carla Yoder.

Carla was already an active seed saver and she had come to the event to learn more about what needed to be done to help get a seed library established in her town. She had also brought some seeds with her to participate in the swap. One seed Carla was offering that was of particular interest to me was a beautiful red and white bean named Yooni's Ennie Bona.

Now, this certainly wasn't the first time that I had encountered a beautiful bean with a unique and interesting name, but it was Carla's research into this particular variety's history that really grabbed my attention. With this jar of beans also came a hand-drawn family tree, tracing this seed back through the Yoder family all the way back to this seed's namesake, Anna Miller, born in Ohio in 1904! I had never seen such a detailed account of a family line, from the perspective of that family's seed. This paper was filled with names, dates and locations that illustrated the direct connection that this bean has with Carla's family history. In my eyes, the family's history and the history of this

SAVING OUR SEEDS: The Practice & Philosophy

variety are one in the same, and Carla Yoder's dedication and success with researching, documenting and preserving this history is the ultimate example of the important work that every seed keeper takes on when they choose to plant, save and share their seeds.

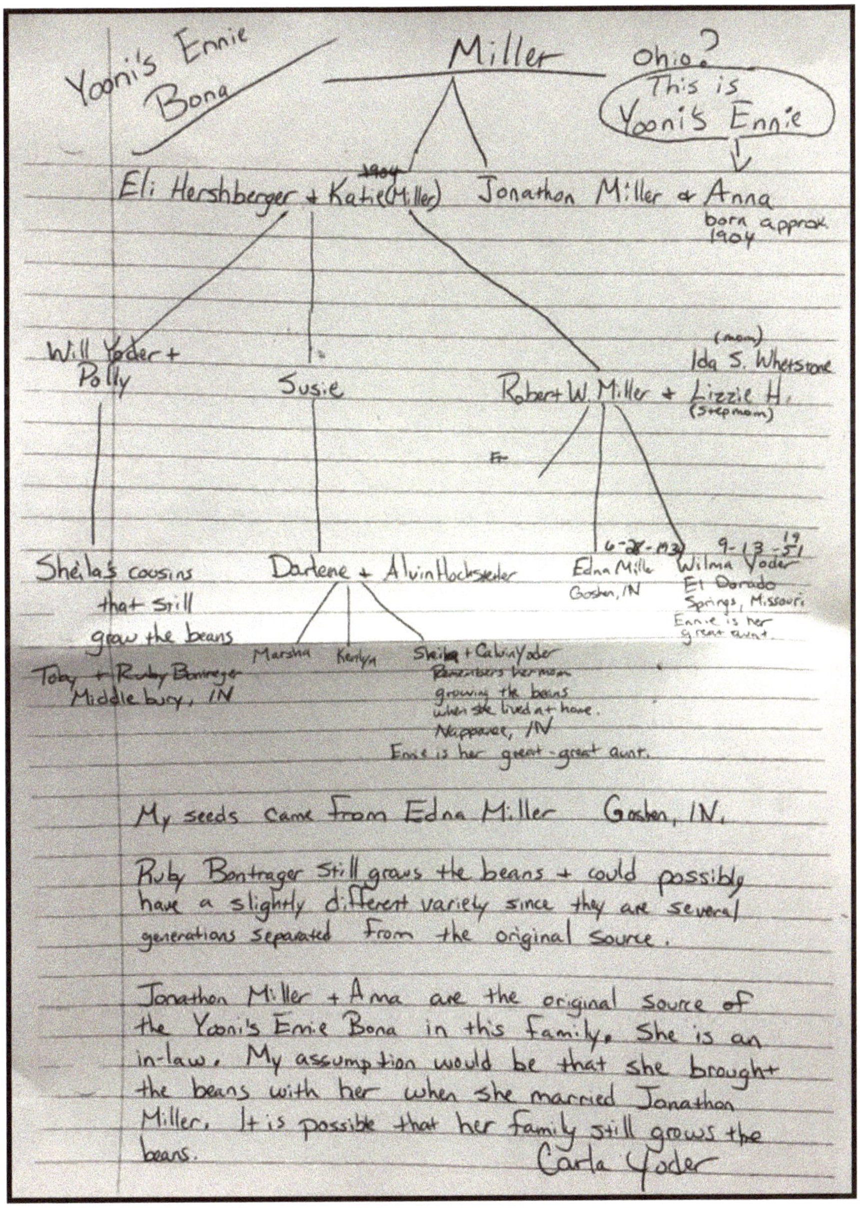

The family tree of Yooni's Ennie Bona

CARROTS

Daucus carota

Family: Apiaceae

Pollination: Insect Pollinated

Isolation: ½ mile – 2 miles

DRY SEED

B BIENNIAL

OVER-WINTERING REQUIRED ...SEE PG 184

Photo by Baker Creek Heirloom Seed Company

CARROTS

Although the first cultivated carrots were white, yellow and even purple, it's the orange taproot that has become the most well known. In the 17th century, Dutch growers are thought to have cultivated the first orange carrots as a tribute to William of Orange, who led the struggle for Dutch independence.

Carrots are certainly a pleasure to grow in the garden, once one overcomes the difficulty in germinating the seeds, but they can prove to be quite tricky to harvest as a seed crop. The greatest challenge facing a seed saver is the concern of cross pollination with wild carrot, also known as Queen Anne's Lace. Since *Daucus carota* is insect pollinated, it can be quite tricky to produce pure seed in areas where this wild cousin is also found growing. For some growers, caging a population of your carrot plants and introducing pollinators into the area is the solution but I have also read of determined seed savers actually hand pollinating their carrot flowers to ensure no crossing with other carrot varieties!

An additional challenge one will face when growing carrot for seed is the plant's need for vernalization. Since the carrot is a biennial crop, it must be overwintered before it will flower. In areas with a mild winter, carrots can be simply overwintered in the ground with a generous application of mulch for insulation. For gardeners with winters that fall below 15°F (-9°C), the plants must be dug up and prepared for winter storage. More information about overwintering your crops can be found on page 184. Once your carrots have been vernalized, planted out and properly isolated for production, they can be left to flower and finish their life cycle. When mature, the seeds of *Daucus carota* are harvested and processed like dry seeds as described on page 182.

CAULIFLOWER

Brassica oleracea

Family: Brassicaceae

Pollination: Insect Pollinated

Isolation: ½ mile

DRY SEED

B BIENNIAL

OVER-WINTERING REQUIRED ...SEE PG 184

Photo by Baker Creek Heirloom Seed Company

CAULIFLOWER

Cauliflower is one member of a plant species that is possibly one of the most diverse food crops in cultivation. Other members of this species include broccoli, collards, Brussels sprouts, cabbage, kohlrabi and kale. *Brassica oleracea* is typically considered a biennial but some of the plants in this species are annuals and will produce seed in the first season they are grown. Cauliflower is typically a biennial and will require overwintering in order to flower and produce seed. More information on this process can be found on page 184.

It is widely believed that modern cauliflower was developed over time from broccoli. The edible parts of the head are called curds and are actually undeveloped florets. They are available in a range of colors including orange, purple, light green and most commonly, white. It's interesting to note that white cauliflower heads are actually produced through blanching. While the plants of this species are all self-incompatible, they will readily cross pollinate amongst themselves and caution must be taken to avoid this.

Since many of these plants are biennial, the grower can enjoy a diverse selection in one season while allowing one variety from the previous year to go to seed. Although the many crops of *B. oleracea* may have various growing needs, they essentially all produce seed in the same manner and can be processed with similar techniques.

Seeds of *Brassica oleracea* plants are produced in small pods call siliques. When the seeds are mature, these siliques will dry and turn a light brown color. Care must be taken when gathering the dried pods as they can easily shatter and the seeds will be lost. Once collected, the pods can easily be flailed or threshed to remove the seeds which can then be easily winnowed or run through screens to remove the debris. More details about threshing and winnowing dry seeds can be found on page 182.

SAVING OUR SEEDS: The Practice & Philosophy

CHARD

Beta vulgaris

Family: Amaranthaceae

Pollination:
Wind Pollinated

Isolation:
800 feet – 1 mile

DRY SEED

B BIENNIAL

OVER-WINTERING REQUIRED ...SEE PG 184

Photo by Baker Creek Heirloom Seed Company

Oriole Orange Swiss Chard

This stunning golden chard selection, named after the beautiful golden Oriole bird, is a gorgeous addition your garden and can also be used in ornamental edible landscaping.

It is commonly believed that *Beta vulgaris* was first domesticated near the Mediterranean more than 2,000 years ago and that it was originally cultivated for its leaves and stalks, similar to the Swiss chard that we know today. It wasn't until around the sixteenth century that the root crop we have come to know as table beets, or beetroot, was developed. It's important for the seed saver to realize that chard and beets are of the same species, as is the well-known sugar beet.

As this species is wind pollinated, great care must be taken to keep your variety isolated to avoid crossing. The pollen produced by these plants is very small and light, so the easiest way to ensure purity is to only allow one variety to flower each year. Swiss chard is a biennial and will need to be over wintered as described on page 184. With some advanced planning and an alternating schedule, you could easily maintain a few different varieties of your favorite beets and chards.

You will want to plan extra space for your second year chard plants as they can grow quite large and will likely also require staking. When the fruits that form on the branches begin to turn brown and dry, they can be collected and cleaned like other dry seeds. More information about processing dry seed crops can be found on page 182.

Hello Dahlia

by Bevin Cohen

Our little farm and homestead, which we have lovingly named Small House Farm, is nestled in the woods on a rarely traveled, dead-end dirt road. Although the name of our farm frequently causes visitors to assume that we must be living in a "tiny house", a portable home on wheels with just enough living space to almost accommodate two adults, two quickly growing children and a rotating selection of housecats, I assure you that this is not the case. We do, in fact, live in a regular sized home, with plenty of space for everyone to be comfortable…on most days. Our two boys keep getting bigger and the house remains the same size, so we'll see how it works out for us in the long run!

While Small House Farm is quietly tucked away in the country, we are not too far from a couple of small towns, each about a twenty minute drive away, one to the east and the other to the west. Although both of the cities are less than an hour apart, they couldn't be more different from one another! Heading west will take us to a small college town, an active and somewhat exciting place to visit when school is in, and a mellow, lazy town during the summer months. If we were to take the road east we would find ourselves in a city with far more conservative values but with money to spend on parks and projects meant to enhance the quality of life and to set this small town apart from the other small towns sprinkled across the central part of the state. In this town one can find art galleries, coffee shops on nearly every corner, expansive community garden spaces and the setting for the story I am about to share: a place known as Dahlia Hill.

This beautiful garden is located, as one could easily surmise from the name, on a hillside. In fact, it is said that this particular hill is the only naturally

SAVING OUR SEEDS: The Practice & Philosophy

occurring hill in the entire city! As the story goes, Dahlia Hill came to be thanks to artist and teacher Charles Breed who moved to the city with his wife Ester in 1950. Sixteen years later, Ester received a gift of dahlia tubers on Mother's Day and Charles was soon obsessed with these beautiful and unique flowers. Within the next twenty five years Charles had amassed a collection of over 1700 tubers!

By 1992, Charles began to plant his dahlia tubers on this hill and named his new garden Dahlia Hill. Six years later, in 1998, the Dahlia Hill Society was formed and incorporated as a non-profit organization that still manages this unique garden space to this day. They now plant over 3,000 tubers from more than 250 different varieties of dahlia, which is quite a breathtaking and colorful sight to see on a warm sunny day as one strolls through the one and a half acre hillside garden.

The Dahlia Hill Society holds an annual fundraiser each spring where the community is invited to visit and purchase tubers from their expansive collection, to bring some of this beauty and wonder to one's home garden. The selection is unlike anything I have seen before, so one spring I made the trip to town to take advantage of this fine opportunity.

While most of the shoppers browsing around the tables were on the hunt for tubers that would produce distinctive flowers shapes with eye-catching colors, I was on a slightly different mission. I was looking for food. The dahlia is native to Mexico where it has a long history of culinary use. It is said that the Aztecs used the sweet potato-like tubers as food and the dahlia is still considered one of the native ingredients in Oaxacan cuisine.

Although each variety of dahlia for sale that day had a full color photograph of its beautiful flower on display above the paper bag on the table holding the tubers available for purchase, I was far more interested in the size of the tubers themselves. As many modern dahlias have been bred for their visible

beauty it seems that the tubers themselves have gotten smaller and smaller. If I was to find a variety that would be useful as a food, I needed to find a dahlia with a substantial tuber size! Unable to locate one, I decided to ask the caretaker of the garden that was working the sales table that day. I waited in line at the table for quite some time as many excited gardeners purchased their new treasures, asked questions or inquired about growing tips. When my turn to approach the table arrived, I simply asked the gentleman working there which variety of dahlia would he recommend for eating and, much to my disappointment, my question was answered with a blank stare and then a short, "Um, I don't.. I don't know."

As it turns out, no one had ever visited the annual Dahlia Hill tuber sale fundraiser looking for varieties best suited for eating, so my question that day was going to go unanswered. But the man working the table was not one to just give up and he helped me sort through the brown paper bags as we hunted for the largest available tubers he had. We figured that size was an important factor in picking a good dahlia tuber meal and that's where we would start this adventure. We were able to find a few decent choices at a very reasonable price (he even threw in a couple for free!) and I headed home from Dahlia Hill a happy man.

The car ride home that day gave me a bit of time to contemplate a few things that day about flowers and beauty and traditional Oaxacan cuisine and, most importantly, food security. At first it seemed strange to me that the caretaker of one of the largest dahlia collections I have even seen seemed to not be aware of this tuber's history and use as a food source. But then again, why would he? Although he may spend his seasons planting, tending and eventually harvesting these dahlias, at the end of the day his dinner likely comes from the grocery store, or perhaps a local restaurant. Most folks these days are too far disconnected from their food and they most likely don't even know where the most basic ingredients on their plates originate. How could this gardener be expected to know how to use these ancient Aztec tubers

that he spends his summers caring for? I couldn't help but wonder what the people at the local soup kitchen just a few blocks down the road might few feel knowing that an entire hill of potentially delicious, and certainly nutritious, food was growing right around the corner from where they spent their afternoons. But this garden wasn't for food, it was for beauty. This garden was a place for people like me, people that don't worry each day where their next meal might come from, to stroll about and enjoy the pretty flowers on a warm and sunny summer afternoon.

As I planted my newly purchased dahlia tubers into the soil at Small House Farm, I had to take a moment to be thankful for the abundance that Mother Earth makes available to me, to realize that understanding the history and stories carried within my food is a blessing that not everyone with their hands in the dirt may realize. It's important to share what we know and how we feel about our food, about our world, and maybe someday we can all enjoy a delicious meal of dahlia tubers together… at our local soup kitchen.

CHICKPEA

These popular legumes are also known as garbanzo beans, a term that became popular in Spain around 1759. It's thought that the Spanish got the word garbanzo from the Basque word "garbantzu", which translates to "dry seed."

The chickpea is an ancient crop, having been first domesticated near present day Syria around 12,000 years ago. It's probably most well-known here in the states as the main ingredient in hummus, but in our house (and around the world) it is a very versatile ingredient in a number of dishes. The type of chickpea most people are familiar with is the Kabuli type, a larger cream colored seed, but there is also the Desi type which is smaller, wrinkled and often brown or black. We have a variety that was gifted to us by Canadian plant breeder Telsing Andrews that is a beautiful mix of colors, including green!

Since *Cicer arietinum* is self-pollinating, the concern for cross pollination is minimal but the easiest way to avoid this is to simply grow only one variety per season. Be sure to plant a good number of chickpeas in your garden as each pod will only contain one or two seeds on average. This will ensure that your harvest is large enough to enjoy in the kitchen and you will still have plenty of seeds for the following year's garden. Chickpea is a dry seed when harvested and is processed with the same technique used for other beans. More information on threshing and winnowing dry seeds can be found on page 182.

SAVING OUR SEEDS: The Practice & Philosophy

COLLARD

Brassica oleracea

Family: Brassicaceae

Pollination:
Insect Pollinated

Isolation:
½ mile

DRY SEED **BIENNIAL**

OVER-WINTERING REQUIRED ...SEE PG 184

Photo by Baker Creek Heirloom Seed Company

Old Timey Blue Collard

Donated to the Seed Saver's Exchange in 1989 by Ralph Blackwell of Alabama. In his donation letter, Ralph described how this variety has been grown by his family for over a hundred years. His mother made a dish similar to sauerkraut from the leaves.

Collards are one member of a plant species that is possibly one of the most diverse food crops in cultivation. Other members of this species include broccoli, kale, Brussels sprouts, cabbage, kohlrabi and cauliflower. *Brassica oleracea* is typically considered a biennial but some of the plants in this species are annuals and will produce seed in the first season they are grown. Collards are a biennial and will require overwintering in order to flower and produce seed. More information on this process can be found on page 184.

It's believed that collards were one of the first forms of domesticated *B. oleracea* which was cultivated near the eastern Mediterranean more than 2000 years ago. Although synonymous with southern cuisine, collards didn't arrive in the Americas until brought here by Africans in the early 1600s. While the plants of this species are all self-incompatible, they will readily cross pollinate amongst themselves and caution must be taken to avoid this.

Since many of these plants are biennial, the grower can enjoy a diverse selection in one season while allowing one variety from the previous year to go to seed. Although the many crops of *B. oleracea* may have various growing needs, they essentially all produce seed in the same manner and can be processed with similar techniques.

Seeds of *Brassica oleracea* plants are produced in small pods call siliques. When the seeds are mature, these siliques will dry and turn a light brown color. Care must be taken when gathering the dried pods as they can easily shatter and the seeds will be lost. Once collected, the pods can easily be flailed or threshed to remove the seeds which can then be easily winnowed or run through screens to remove the debris. More details about threshing and winnowing dry seeds can be found on page 182.

SAVING OUR SEEDS: The Practice & Philosophy

CORN

Zea mays

Family: Poaceae

Pollination:
Wind Pollinated

Isolation:
½ mile – 1 mile

DRY SEED   ANNUAL

Photo by Baker Creek Heirloom Seed Company

Strawberry Popcorn

The cute little ears of this corn look just like big strawberries and are great for fall decorations or making delicious popcorn. This popcorn is a fun garden crop for kids of all ages.

CORN

I could wax poetic about corn all night long, but unfortunately not everyone finds this beautiful plant to be as romantic as I do. In fact, in America alone, nearly 90 million acres of corn are grown annually and around 75% of that crop is harvested for ethanol production and animal feed. Truly, only a fraction of the millions of acres of monoculture corn fields we see growing throughout Iowa and most of the Midwest is for human consumption and a great majority of that is in the form of high fructose corn syrup.

It's in community gardens, backyards and urban farms where you can still find people lovingly tending these beautiful and ancient grains every summer and late into the fall, harvesting the gloriously colored ears for corn meal, hominy and popped corn.

Since *Zea mays* is wind pollinated, the easiest method to ensure pure seeds is to simply grow only one variety per season. It's important to note that this species suffers greatly from what is called "inbreeding depression", so one must be sure to harvest seed from a minimum of 100 plants. If you must grow more than one variety in a season or if you find your garden is to close to a neighbor's corn patch, hand pollination is the key to avoiding crossing and maintaining varietal purity. More information about hand pollination can be found on page 172.

Mature corn seed is left on the plant as long as possible to dry down, then harvested and shucked and allowed to dry further indoors, away from the elements. It can later be removed from the cob by hand or with a mechanical corn sheller. At this point it is easy to clean by screening or winnowing. More details about winnowing your seed can be found on page 182.

SAVING OUR SEEDS: The Practice & Philosophy

SEEDKEEPER'S TALE

RAFAEL MIER
MEXICO CITY, MEXICO
FOUNDER, TORTILLA FOUNDATION

My entry into the world of corn was totally unexpected. I had organized a small party for my family and some friends to celebrate my birthday. That day, we ate some delicious corn tortillas that were made of the corn I had grown on my farm in Valle de Bravo. The tortillas were made the traditional way, with corn nixtamalized the ancestral way and with the tortillas shaped by hand at the moment of cooking.

In the middle of the celebration, some of the guests made comments praising the flavor and quality of the tortillas. The conversation became a discussion about the process of making a corn tortilla. I was surprised to discover that most of the partygoers were completely unaware of the nature of this food that is so important in Mexico's culture. Realizing this made me reflect on the disconnect that we, who are residents of Mexico's big cities, have with respect to the origin and the history of our foods. The corn tortilla is the food that is most consumed in Mexico and is the principal source of calories and proteins for Mexico's people. The tortilla is one of the most important cultural and culinary elements for our culture. It has been eaten for around 2,000 years and during all of that time, it has suffered very few changes.

Photo by Baker Creek Heirloom Seed Company

Even though the tortilla is a hugely important food for us Mexicans, it is currently experiencing serious problems. In the last 30 years, the per capita consumption of corn tortillas has fallen 40%, and to make the problem worse, the corn tortillas that we are eating now are of the worst quality. We must also take note that during this same time period, our health has been seriously affected by a deficient diet. Proof of that is the growth of a population that is overweight, obese and diabetic, placing Mexico among the top spots with these problems on an international level. Nevertheless, even though the data about the decline of the corn tortilla is so alarming, it has not become the subject of close attention either in public politics or any societal initiative.

Beginning several years ago, I felt the need to speak about the deterioration of food in Mexico. I had written several articles about this and had sent them to many newspapers and magazines without successful publication. The experience that I had upon realizing the need for information about the tortilla made me understand that in order to achieve a change in what we Mexicans eat, the change had to come about beginning with its essential food, the corn tortilla. Deep within myself I thought: if we don't make Mexico take into account the quality of its corn tortilla, it will be with great difficulty that we realize the quality of other foods of less relevance. That day I proposed to achieve what I believe; that Mexico must reflect on the importance that the corn tortilla has for its culture, its diet and the social and economic development of our country.

My first reaction was to write something about the corn tortilla, but as I had already endured a negative experience trying to publish my articles, it occurred to me that I should open a Facebook group, which I called Tortilla de Maíz Mexicana and opened in November of 2015. I thought also of the differences that our tortilla has with respect to the tortillas that are eaten in Guatemala and other countries of Central America. I started publishing brief messages and images every day about the process of making a tortilla, the diversity of native corns that exist in Mexico and other topics related to corn and the tortilla. At the beginning, ten or twenty friends were following the group on Facebook but little by little the number grew. Today, we have more than 365,000 followers.

In February of 2016, I decided to formalize Tortilla de Maiz Mexicana and we are now a legal non-profit organization whose objective is to promote the culture and consumption of native corns, to recuperate a quality tortilla and to promote and protect the biodiversity of native corns. Later we decided to change the name of this initiative to Fundación Tortilla (the Tortilla Foundation).

We are a small organization, but we have achieved important goals. During these last three years, we have given more than 130 conferences in universities, schools, regional fairs and markets for producers. We have continued the work of creating content specializing in corn and tortillas, spreading the message through our Facebook page. Recently we have also started taking our message to Instagram. We have filmed short videos that document the culture of corn and tortillas, videos that have generated millions of views on social networks. By the same token, we have collaborated with other communications media to foment the spread of knowledge about corn and the tortilla. We are also working to recover our ancestral popcorns, they are the oldest of all the corn in the world and which are today in danger of extinction.

Right now, we are pushing for changes in the norms that regulate the making and selling of corn tortillas in Mexico. We seek to achieve a fair market that makes distinctions between types and qualities of corn tortillas made commercially in Mexico. We want a regulation that restricts chemical additives, prohibits the use of artificial coloring and which protects the consumer from the consumption of genetically modified corn and the agricultural chemical residues that accompany them.

Without a doubt, these last four years have been the most gratifying of my life. I have had the opportunity to be part of the wonderful world of corn and it never ceases to surprise me with its grandeur and the importance it plays in our culture.

COWPEA

Vigna unguiculata

Family: Fabaceae

Pollination: Self-pollinating though occasionally pollinated by insects

Isolation: 10-20 feet

DRY SEED

A ANNUAL

Photo by Baker Creek Heirloom Seed Company

Black Crowder

The crowder pea gets its name from the way its starchy peas crowd themselves in the pod. It's said that this is the best tasting, most prolific and drought resistant cowpea available. The peas themselves have a deep purple color when first shelled which then turn black when dried. Originally introduced in 1907.

COWPEA

The cowpea, or southern pea, is nearly synonymous within southern cuisine, but we can trace the origin of this tasty little legume back to the continent of Africa. It is believed that *Vigna unguiculata* first made its way across the Atlantic in the 17th century, transported by the Spanish, and ended up in the United States a century later, being brought over from Africa through slave trade.

One of the most well-known varieties of cowpea is 'Black Eyed Pea', an off-white seed with a black marking located on the hilum but there are hundreds of cultivars of many diverse colors including black, red, brown, purple and cream. While cowpeas are typically grown for their mature dry seed, there are types that were bred in Asia that are grown for their long and tender fresh pods. These varieties are known as yard long beans.

Regardless of whether one plans to grow their southern peas for dry seed or fresh eating, harvesting the crop for seed is handled the same way as any other dry bean grown in the garden. More information on harvesting, threshing and winnowing dry seed can be found on page 182.

SAVING OUR SEEDS: The Practice & Philosophy

SEEDKEEPER'S TALE

ANGIE LAVEZZO
NORTH CAROLINA, USA
MANAGER, SOW TRUE SEEDS

Everyone loves an underdog. The underdog archetype appeals to our collective humanity, the belief that we are all inherently good and kind and just trying to do the best we can on this big ol' ball orbiting the sun. I think this is why I became a seed saver, a steward of old varieties I was afraid would be lost and forgotten. Seeds are underdogs, full of mystery and fragility and if I could nominate an underdog in a family of underdogs, it would be the lowly cowpea.

Let me clarify that I don't think cowpeas are lowly at all and I honestly LOVE cowpeas. I was twenty when I grew my first crop and admit that I had no real idea what I was growing. I planted the Pinkeye Purple Hull variety (I thought it sounded pretty) in late March because I thought I was growing a funny looking English-type pea. Living at the time in Northern Virginia, it just happened to be warm enough to germinate and, as the weather improved, I received a very prolific harvest of what was clearly not a traditional pea. I finally broke out a gardening book and looked up the species, *Vigna unguiculata,* and thus was born a long love affair with the cowpea.

SAVING OUR SEEDS: The Practice and Philosophy

Photo by Baker Creek Heirloom Seed Company

Cowpeas are really just misunderstood. Their long history as animal feed has left them far from the forefront of culinary stardom. I have gotten more than a few funny looks when I admit it's my favorite. Even after I moved to North Carolina, where the cowpea is a common ingredient in many classic dishes, my declaration of love is most often met with polite Southern curiosity. What is it about the cowpea that inspires such indifference? Why is it ignored by such a large portion of gardeners? Where is the love for the Southern Pea?

Well, don't worry y'all. I have enough love for this veggie that my enthusiasm will make you grow your next crop with childlike anticipation.

As anyone interested in heirloom seeds knows, the more popular legume, *Phaseolus vulgaris,* has folks that are dedicated to hunting down and preserving its vast genetic diversity. This is a fun and almost endless hunt, because the amount of germplasm out there waiting to be rediscovered, and in need of saving, is awe inspiring. While there are not as many cultivars in *Vigna unguiculata* as in *Phaseolus vulgaris*, there are still hundreds of varieties out there to sample. The 2019 Seed Savers Exchange yearbook alone lists 154 different varieties, representing different colors, shapes, sizes, tastes, and textures.

So why grow cowpeas? Why not just grow some of the endless varieties of snap or shell beans? Well this is the best part: cowpeas are easy! Easy to grow, easy to harvest, easy to shell, easy to cook. They laugh in the face of mid-summer heat and after flowers start setting, rarely need irrigation. When my snap beans and greasy beans are withering in the sun or struggling to survive the onslaught of bean beetles, my cowpeas are chuckling and having a cocktail. How promising is this for growing in an ever hotter planet? When I'm hunting through bean vines to find their treasures, my cowpeas are reaching their pods above their foliage for an easy harvest. When I'm cranky from stringing my half runners and shelling my soup beans, the cowpeas let me squish their shell between my fingers so the tender peas just pop out into my bowl like edamame. Their smaller size cooks up quicker than their larger cousins, for a hearty, protein-packed (averaging almost 25% protein!) meal. What are you going to do after reading my ode to *Vigna unguiculata*? You're gonna go cook some cowpeas!

Try this easy recipe:

Soak 1 ½ cups of dried cowpeas overnight.
Sauté at least one large onion and a few cloves of garlic in a stew pot.
Add your soaked peas, 3 ½ cups water, 2 cups chopped tomato (canned works too!), 1 ½ tsp each of salt, cumin, and paprika, 1 tsp ginger, and a pinch or two of cayenne, to your taste.
Cook for about an hour and a half until peas are tender,
adding more water if needed to keep a nice gravy consistency going.
Add a handful of fresh cilantro (if you tolerate it) at the end
and serve it over rice.

This is my favorite way to eat cowpeas. It's inexpensive, extremely satisfying, and so very good for your body and soul. It's super adaptable too; add whatever you want to it! The tomato gravy is key because it'll seep down into the rice, balance out the meaty texture of the cowpeas, and is just knee slappin' good.

What are you going to do after you eat the best meal of your life? Well, I hope you're going to hunt down some interesting cowpea varieties and add them to your garden rotation. They deserve a row or two, and the seeds are easy to save too. We need to take this work seriously, because we have lost so much genetic diversity over the last century. Our seeds are our history and worth our allegiance: especially the underdogs.

Straight Eight Cucumber

This tasty cucumber gets its name from its consistent production of perfectly straight, 8-inch fruits. First introduced by Ferry-Morse in 1935, this variety was also an All American winner that same year.

Cucumbers are one of the crops that always make me think about summer. As a vendor at the local farmers market, I have watched eager customers hauling away bushel baskets of these juicy fruits with plans to take them home to wash, slice and pickle; a wonderful way to enjoy the pleasures of the harvest all winter long. Cucumbers are easy to grow, even in areas with a shorter season, and they don't take up nearly as much space as other vining members of the Cucurbitaceae family. We have had great success trellising cucumbers in our home garden; this technique keeps the fruits out of the dirt and it's much easier on your back come harvest time!

While growing *Cucumis sativus* for seed is a relatively straight forward process, there are a couple of important details to take into consideration. First, since cucumbers are insect pollinated, they must either be isolated from other varieties of the same species or hand pollinated in order to avoid crossing. The specifics of hand pollination can be found on page 172.

Secondly, cucumbers must be left growing on the plant to full maturity in order to harvest any viable seed. The point in a cucumber's life where it is considered "market ready" is not the same as the plant's botanical maturity. Don't worry, your cucumber will let you know when it's ready to harvest for seed by noticeably changing color. Typically, you will see your fruits turning yellow to indicate its readiness, at which point you will harvest the mature cucumbers and process as wet seeds. See page 176 for step by step instructions on harvesting and cleaning wet seeds.

EGGPLANT

Solanum melongena

Family: Solanaceae

Pollination:
Self-pollinating as well as insect pollinated

Isolation:
300-1,500 feet

WET SEED

A ANNUAL

Photo by Baker Creek Heirloom Seed Company

EGGPLANT

Eggplants are another one of those garden crops that I wish I would have taken more time to appreciate as a child. They are stunning in appearance and versatile in the kitchen. As an adult, I can barely get enough eggplant onto my plate and I am so thankful for the incredible diversity offered by this amazing fruit! There seems to be many different thoughts on the origin of *Solanum melongena* with some scholars believing that it was first domesticated in India, while others believe that domestication took place in China. There are written records from both of these countries mentioning eggplant that are over 2000 years old. There are also wild relatives of eggplant that are still found in both Africa and Asia.

While eggplant is self-pollinating, insects are certainly drawn to its flowers, which can easily cause crossing between varieties. There are a few solutions to this challenge including isolation, bagging of blossoms and simply growing only one variety each season. More information about blossom bagging can be found on page 173. Another way around this cross pollination concern is growing an eggplant from a different species. We have grown, and loved, a variety of *Solanum aethiopicum* which has delicious and beautiful scarlet colored fruits. This way we can enjoy the diversity of two eggplant varieties with no concern for crossing, guaranteeing us a harvest of pure, true to type seeds. The technique for harvesting seeds from both of these species is the same.

Be sure to let your eggplants fully mature past the eating stage if you plan to harvest viable seed. You will know your fruits are mature when they begin to change color to yellow or brown and in the case of *S. aethiopicum*, they will turn orange. The seeds will be processed as wet seeds, a technique which is described on page 176. A trick to simplify the extraction process is to cut the mature fruit into small cubes and place them in a food processor with a small amount of water, using a dough blade and a couple quick pulses to easily remove the seeds from the fruit pulp.

SEEDKEEPER'S TALE

MEHMET ÖZTAN
WEST VIRGINIA, USA
OWNER, TWO SEEDS IN A POD HEIRLOOM SEED COMPANY

I, once, was an engineer. Then one spring, years ago, I planted seeds in the backyard of our new home in Tampa, Florida and after that point in time, I could never go back from loving the act of gardening.

I am a Turkish seed saver and farmer who was born and raised in Turkey by my lovely parents, who always provided the freshest food for our family. My father was a professor of forestry, who mentored some of the finest experts in the field, his brother was a renowned landscape architect and they both loved sharing stories about trees and flowers at the dinner table. Alas, I grew up in the city of Ankara, in an apartment building all my teen life. Yes, there were cherry and mulberry trees in our backyard which I had the extreme joy of climbing with my friends in summer but I never grew vegetables until after I moved to Tampa.

I vividly remember seeing produce grown from Turkey's signature heirloom varieties such as *Sürme* watermelon, *Ayaş* tomato, *Çengelköy* cucumber, *Adapazarı* squash, İspir dry bean and many others sold specifically with the varietal names, at almost all grocer's shops (tr. *manav*) and farmer's markets (tr. *pazar*) nationwide, well into the 2000s.

SAVING OUR SEEDS: The Practice and Philosophy

Photo by Baker Creek Heirloom Seed Company

Since after I came to the U.S. as a Ph.D. student in 2006 at Michigan State University, one thing I was desperately seeking was the flavors and variety of all the vegetables from my childhood. In 2010, I moved to Tampa, following my then soon-to-be wife, Amy Thompson. That summer I started a vegetable garden in our new home's backyard, while I was trying to finish my dissertation remotely, which not only helped me to relieve stress but also opened the door to the realm of so many flavor notes, fruit colors, shapes and plant types for me. Soon after, I started searching for Turkish heirloom seeds through seed swaps with seed savers, Seed Savers Exchange's seed collection and various seed companies' catalogs. Then I found out about the struggles of small farmers and how multi-national biotech companies control the national and global seed markets. Saving and preserving seeds to protect biodiversity and my cultural values eventually became a passion for me. A few years later, with the encouragement from Amy, I quit my engineering career to turn my passion into a full-time job by founding our heirloom seed company, Two Seeds in a Pod. Soon after, our backyard wouldn't be enough for my seed saving efforts and I would expand into bigger gardens and eventually a 3-acre field.

Among all vegetables Amy and I were eating from our garden, eggplant (tr. *patlıcan*, also called *balcan* or *badılcan* in different dialects) quickly became one of our favorites, as they resembled the ones I remember eating in Turkey and they had much better flavor than the ones sold at grocery stores here in the United States. Eggplant is a culturally very important crop for me as well. Although it is known to be native to South Asia, Anatolia (also known as Asian Turkey and Asia Minor) is one of the oldest places that it has been cultivated. There are many Turkish/Anatolian eggplant varieties as well as dishes made from it, such as *Karnıyarık* (eng. eggplant stuffed with ground beef) to *Kuru Patlıcan Dolması* (eng. stuffed, dehydrated eggplant) to Şakşuka (eng. fried eggplant served with plain yogurt and tomato sauce) to İmambayıldı (eng. eggplant stuffed with vegetables) to *Yoğurtlu Patlıcan Salatası* (eng. grilled eggplant salad served with garlic and plain yogurt). In the last ten years, I collected, grew and shared seeds for many Turkish eggplant varieties that have exquisite flavor and impressive culinary values, including, *Kemer*, *Aydın Siyahı*, *Pala*, *Antep*, *Topan*, *Alacalı Manisa*, *Yamula*, *Halep Karası* (a shared cultural gem of Syria and Turkey), and one that I had the privilege of naming and that I rank equally with *Halep Karası* in terms of flavor, *Muğla Yılanı* (eng. Snake of Mugla).

After living, gardening and farming for about 10 years in Florida, in summer 2018, we moved from Florida to beautiful West Virginia. I now work at West Virginia University in fall and spring semesters as a Service Assistant Professor in the department of Geology & Geography, but not as an engineer, as a seed steward who promotes heirloom seeds, seed sovereignty, seed security and sustainable farming practices.

In early 2019, we moved our seed company to Reedsville, West Virginia, where I continue my adventure with Turkish seeds and the wonderful heirloom varieties of Appalachia and West Virginia on our 6-acre seed production and research farm. As of summer 2019, I will be growing, eating and harvesting seeds for 10 eggplant varieties and I look forward to growing many more of them in the years to come.

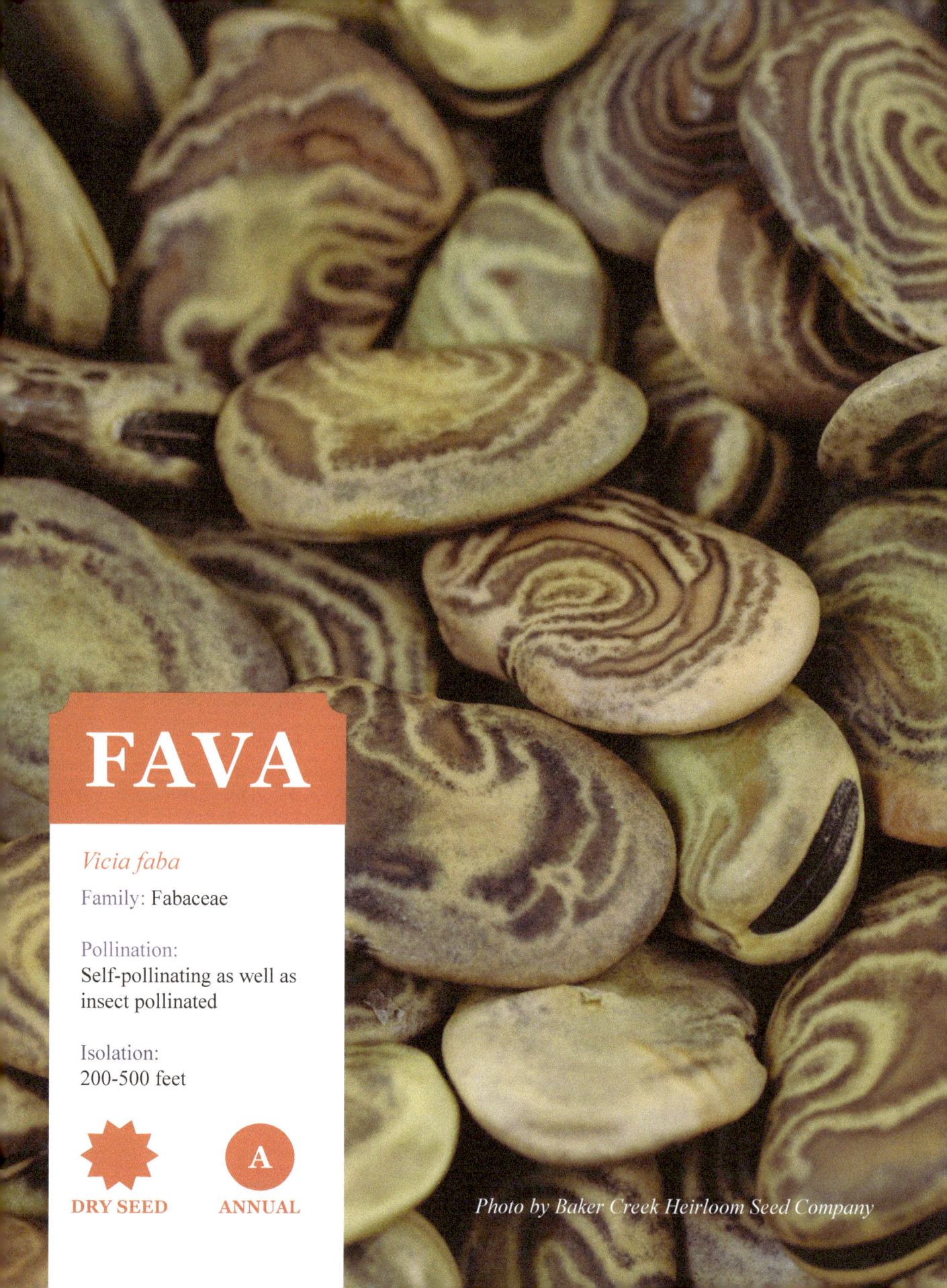

FAVA

Vicia faba

Family: Fabaceae

Pollination:
Self-pollinating as well as insect pollinated

Isolation:
200-500 feet

DRY SEED

A ANNUAL

Photo by Baker Creek Heirloom Seed Company

Habas Qhelka Fava

This gorgeous variety is also known simply as 'Fingerprint'. They are a very special local type in the highlands of Peru and were considered sacred and adored by the Inca.

Fava beans are a truly ancient legume having been utilized in the area of modern day Syria over 12,000 years ago although the large seeded phenotype that we are most familiar with today was first documented by archaeological evidence dating back only 2000 years. *Vicia faba* enjoys cool weather and will do well when planted early in northern climates or planted in the fall and overwintered in areas that have winters that don't regularly reach temperatures lower than 15°F. (-9°C)

Fava beans are harvested and processed like any other bean, being a dry seed and requiring threshing and winnowing to clean the bean for food or storage. More details on threshing and winnowing can be found on page 182.

While this species is self-pollinating, the large flowers are very attractive to insects and crossing between varieties is quite common. The easiest way to avoid this is through isolation or by simply growing only one variety per year. While this species will self-pollinate and bagging or caging the plants to avoid cross pollination is an option, there have been a number of studies that have shown that *Vicia faba* yields are notably larger and earlier when the favas are left to cross with surrounding plants. This would lead one to then conclude that growing only one variety, or ensuring proper isolation between varieties, is the most productive and sensible decision. For growers interested in bagging their fava bean blossoms, more information can be found on page 173.

GROUND CHERRY

Physalis grisea

Family: Solanaceae

Pollination:
Self-pollinating as well as insect pollinated

Isolation:
300-1500 feet

WET SEED

A ANNUAL

Photo by Baker Creek Heirloom Seed Company

Cape Gooseberry Ground Cherry

An excellent choice for making old-fashioned preserves or for use in Mexican cuisine. Harvest this delicacy from its paper-like husk all summer long. The leaves on these very ornamental plants are extra furry and soft.

The delicious ground cherry has quickly become a favorite around our homestead, for the children and adults alike! Incredibly productive, just a few plants will produce hundreds of fruits throughout a season. And with each tasty ground cherry holding a good amount of seed, it's easy to collect plenty for next year's garden and to share with anyone interested.

The fruits of *Physalis grisea* grow inside of a papery husk and will turn from green to yellow when ripe. When fully mature, they will drop from the plant onto the ground and this is where you will collect them. The seeds are ready to be harvested when the ground cherries are ready to be eaten; just remove them from their papery shell and quickly run them in the blender to remove the seeds from the pulp. They are easily processed as wet seeds and more information on this can be found on page 176.

While *Physalis grisea* is self-pollinating, insects do find the flowers attractive and will cross pollinate your varieties if they are not properly isolated. Bagging the individual flowers can prove to be quite difficult if not impossible, so using row cover may be the best option to ensure seed purity. If this is not an option in your garden, consider growing only one variety from this species to avoid cross pollination.

Historic Gardens, Frozen in Time
By Bevin Cohen

My first experience with a historic garden was likely at the Chippewa Nature Center, a nature preserve located just outside of Midland, Michigan. At the CNC, they have a historic homestead cabin complete with a schoolhouse and sugar shack, as well as a barn, root cellar and garden spaces. It's a beautiful place that highlights what life was like for many people that lived in the area in the 1870's. If you've read my book *From Our Seeds & Their Keepers,* you'll be familiar with the location as I tell the story of when I first met the garden caretaker and he shared some of his very special bean seeds with me. My family used to visit this historic homestead nearly every week and I was simply fascinated by the antiquity of everything there, especially the history of the garden. This is where I learned about heirloom vegetables and traditional gardening techniques. This is also where I first learned about the important practice of seed saving. This historic garden was like a magical place to me, a snapshot of a moment in time, perfectly preserved for everyone to enjoy. The caretakers of this garden worked very hard to make sure that everything they grew resembled, as closely as possible, exactly what was grown in gardens nearly 150 years ago!

Over the past few years, I have had the opportunity to visit a number of other historic gardens and I've also chatted for hours with caretakers of gardens that I haven't had the privilege of visiting yet. My work with heirloom and heritage seeds has helped me to understand the significance of these historic sites and each garden manager's meticulous attention to detail and accuracy is absolutely commendable. When a person visits one of these gardens, it's almost as if you are taken back to another time with fruits, vegetables and flowers offering colors and flavors that simply cannot be found in the

modern grocery store. Each garden is like a photograph, capturing that perfect moment to be repeated over and again for each lucky visitor to enjoy.

I know that my gardens are never like that. No, my gardens are always in a state of flux. Of course they are always changing throughout a season. The battle with weeds never ends and the vines continue to expand their reach until they have ended up far beyond where I originally planted their seeds. This is normal; plants grow and gardens change. What I'm referring to is an annual flux. While I may sometimes grow the same variety from year to year, I am working hard to adapt that plant to my particular microclimate and even my personal gardening style. I accomplish this through seed saving. When one saves seeds from their gardens, they have in essence become a plant breeder. By selecting preferred traits or best performing plants and choosing to harvest seeds from these specific specimens, a grower can adapt his varieties over time to perform better, produce more and even be more disease resistant.

The practice of saving one's seeds has seen quite the resurgence in popularity in the past few years. More gardeners are discovering this important work, as well as its benefits, and are taking the time to learn these skills once again. Just a few generations ago nearly everyone that grew food understood the importance of saving one's seeds. Even through casual observation, any gardener could tell that plants grown from saved seed performed better than plants grown from commercial seed. These purchased seeds weren't adapted to your soil and climate, the fruits they produced weren't quite as tasty and nutritious as the ones grown from the family's seeds. I've met growers who have continued to grow and pass down family heirloom seeds for well over a hundred years and every one of these gardeners will tell you that their family variety is far superior to anything that can be purchased from the seed dealers today. While part of that opinion may be purely sentimental, these folks have a point. Each season they have selected seed from the best performing and most delicious produce from

their gardens and each season they have grown better and more delicious produce. This is something every one of us can do in our own gardens simply by saving our seeds! This is something that has been done in gardens since agriculture began.

Saving seeds was certainly a necessity for the original caretakers of these historic gardens that I enjoy visiting every year. When first established, these gardens weren't meant to be tourist destinations or educational preserves. Their purpose was food production. Because of this, we know that the original gardeners were actively working to adapt and improve the varieties that they grew to ensure a plentiful supply of produce to last them throughout the year, far beyond just the growing season. A bountiful harvest was critical to survival.

When we visit these historic gardens, which we should as often as possible to enjoy their beauty and to glean their knowledge, it's important to realize that what we see in their fields only represents a mere fraction of time in the lives of the people that originally cared for these spaces. These gardeners and farmers and the crops they maintained were constantly in a state of flux, of forward momentum, working together to evolve and adapt and to grow; always in motion just like Nature herself, and one of the most powerful techniques they utilized was saving and sharing their seeds.

Photo by Sarah Tomac, Tomac Pumpkins

KALE

Kale is one member of a plant species that is possibly one of the most diverse food crops in cultivation. Other members of this species include broccoli, collards, Brussels sprouts, cabbage, kohlrabi and cauliflower. *Brassica oleracea* is typically considered a biennial but some of the plants in this species are annuals and will produce seed in the first season they are grown. Kale is typically a biennial and will require overwintering in order to flower and produce seed. More information on this process can be found on page 184.

Along with collards, kale is considered to be one of the earliest domesticated forms of *B. oleracea*. There is quite a bit of diversity within kale varieties with some ranging from light green to dark purple and leaf types everywhere between smooth to deeply savoyed. Russian kale varieties are actually from the species *Brassica napus* and will not cross with *B. oleracea* kale types.

While the plants of this species are all self-incompatible, they will readily cross pollinate amongst themselves and caution must be taken to avoid this. Since many of these plants are biennial, the grower can enjoy a diverse selection in one season while allowing one variety from the previous year to go to seed. Although the many crops of *B. oleracea* may have various growing needs, they essentially all produce seed in the same manner and can be processed with similar techniques.

Seeds of *Brassica oleracea* plants are produced in small pods call siliques. When the seeds are mature these siliques will dry and turn a light brown color. Care must be taken when gathering the dried pods as they can easily shatter and the seeds will be lost. Once collected, the pods can easily be flailed or threshed to remove the seeds which can then be easily winnowed or run through screens to remove the debris. More details about threshing and winnowing dry seeds can be found on page 182.

SAVING OUR SEEDS: The Practice & Philosophy

KALE, RUSSIAN

Brassica napus

Family: Brassicaceae

Pollination: Insect Pollinated

Isolation: ½ mile

DRY SEED

B BIENNIAL

OVER-WINTERING REQUIRED ...SEE PG 184

Photo by Baker Creek Heirloom Seed Company

Red or Ragged Jack and Scarlet Russian Kale

A pre-1885 heirloom variety, originally from Siberia but introduced to Canada by Russian traders. These beautiful plants boast very tender oak-type leaves and mild flavor.

One of the most well-known varieties of this species of kale is the popular 'Red Russian" which is a very cold hardy type that seems able to sustain even the most brutal winter weather. While *Brassica napus* is also the same species as the rutabaga, and handled in the same way for seed production, it is a different species than many other kale varieties such as 'Dwarf Blue' and 'Lacinato' which is *B. oleraceae*. These two species are not compatible and the gardener should have no worries of cross pollination between them.

While the exact history of Russian kale is uncertain, it is believed that rutabagas were grown in Sweden around 1400. Unlike many of their cousins in the Brassicaceae family, Russian kale is self-compatible and therefore can be grown in small populations and still produce plenty of seed for the grower. While some plants of this species are annuals, Russian kale is a biennial and will need to be vernalized before seed production can occur. More information on overwintering your garden plants can be found on page 184.

The following year, when your plants have gone to seed, the siliques (seed pods) will dry and turn brown when the time to harvest has arrived. Harvesting and processing kale seeds is a simple task, but caution must be taken to avoid shattering the seed pods while gathering them. Once properly threshed, the seed harvest can then be winnowed or screened to remove any debris. More specifics on threshing and winnowing dry seeds can be found on page 182.

It is important to note that Russian kale is the same species as common canola, or rapeseed, and cross pollination can occur between the two. Since this species is pollinated via insect, an isolation distance of ½ mile is recommended.

SAVING OUR SEEDS: The Practice & Philosophy

KOHLRABI

Brassica oleracea

Family: Brassicaceae

Pollination:
Insect Pollinated

Isolation:
½ mile

DRY SEED

B BIENNIAL

OVER-WINTERING REQUIRED ...SEE PG 184

Photo by Baker Creek Heirloom Seed Company

Early Purple Vienna Kohlrabi

Also known as Di Vienna Violetto, this beautiful kohlrabi predates 1860 and is presumably from Austria. An attractive and interesting heirloom addition for any garden.

Kohlrabi is one member of a plant species that is possibly one of the most diverse food crops in cultivation. Other members of this species include broccoli, collards, Brussels sprouts, cabbage, cauliflower and kale. *Brassica oleracea* is typically considered a biennial but some of the plants in this species are annuals and will produce seed in the first season they are grown. Kohlrabi is a biennial and will require overwintering in order to flower and produce seed. More information on this process can be found on page 184.

I have always enjoyed the appearance of kohlrabi. While the enlarged stem of some varieties is green and others have been developed for a purple color, to me, a freshly harvested kohlrabi resembles some sort of alien spaceship; it is truly an unusual variation of this diverse species of food crops!

While the plants of this species are all self-incompatible, they will readily cross pollinate amongst themselves and caution must be taken to avoid this. Since many of these plants are biennial, the grower can enjoy a diverse selection in one season while allowing one variety from the previous year to go to seed. Although the many crops of *B. oleracea* may have various growing needs, they essentially all produce seed in the same manner and can be processed with similar techniques.

Seeds of *Brassica oleracea* plants are produced in small pods call siliques. When the seeds are mature these siliques will dry and turn a light brown color. Care must be taken when gathering the dried pods as they can easily shatter and the seeds will be lost. Once collected, the pods can easily be flailed or threshed to remove the seeds which can then be easily winnowed or run through screens to remove the debris. More details about threshing and winnowing dry seeds can be found on page 182.

SAVING OUR SEEDS: The Practice & Philosophy

LENTIL

Lens culinaris

Family: Fabaceae

Pollination:
Self-pollinating though occasionally pollinated by insects

Isolation:
10-20 feet

DRY SEED

A ANNUAL

Photo by Baker Creek Heirloom Seed Company

Umbrian Orange Lentils

Some of the most prestigious lentils in the world come from France and Italy. These Italian lentils are unique because of their small size and delightful flavor. They are traditionally grown in two small areas around Castelluccio and Colfiorito. In Italy, New Years day is always celebrated with lentils, because they have come to signify prosperity.

It is widely believed that lentils are the oldest of the domesticated legumes, having been grown as a food crop in the area of modern day Syria over 13,000 years ago. *Lens culinaris* is a cool season crop that will do well when planted around the same time as garden peas in northern climates. Lentils are typically tan or yellow but there are varieties with seeds colored green, black, gray and even pink.

These are small statured, self-supported plants with small, self-pollinating flowers. Crossing between varieties is quite rare as insects are almost never interested and the flowers pollinate themselves before even opening. Lentils are harvested as dry seeds and the pods will typically contain two seeds each. Plan to harvest the lentil seed pods before they are completely dry as they will shatter in the garden and the small seeds will be lost. More information on threshing and winnowing dry seeds can be found on page 182.

LETTUCE

Lactuca sativa

Family: Asteraceae

Pollination: Self-pollinating

Isolation: 10-20 feet

DRY SEED

A ANNUAL

Photo by Baker Creek Heirloom Seed Company

LETTUCE

Lettuce all take a moment to enjoy this garden favorite... Silly puns aside, lettuce is surely one of the most popular vegetables grown in home gardens and it is also one of the easiest to grow for seed! It's a self-pollinating species that is almost never a concern for crossing, so minimal isolation is needed to maintain genetic purity.

Lactuca sativa is a very diverse species with varieties that grow as loose-leaf, some with loosely formed heads and also the tightly headed iceberg types. The warmer, longer days of summer trigger lettuce plants to bolt and begin their seed production cycle. While most gardeners are disappointed to see their lettuce plants flower, as it means the leaves will now be bitter and less palatable, as a seed saver this is always an exciting time in the garden! Each composite flower is actually made up of 15-20 individual flowers and each of those will produce a seed. This means that an individual lettuce plant can produce thousands of seeds during its forty day flowering period!

As the seeds reach the mature stage and prepare to be dispersed by the wind, they can be gathered by hand or they can also be bagged if one is planning to gather quite a bit of seed. Lettuce seeds are best cleaned with a combination of screening and winnowing which is described in detail on pages 182.
A quick tip for seed savers that grow head lettuce; when your lettuce has reached market maturity, cut a small 'X' into the top of the head to help the flower stalk properly emerge. Be careful not to cut too deeply and injure the stem.

SAVING OUR SEEDS: The Practice & Philosophy

How Diversity Fills Our Plate

By Bevin Cohen

• •

As I sit here at the kitchen table, staring out my window at the wet, muddy and unseasonable weather we are having this winter in Michigan, I can't help but let my memories take me back to times when the sun beat down from overhead and the soothing notes of bird songs filled the air. Yes, my mind is happy to take this journey through the world of dreams, floating its way backwards in time to the bright green grass of the Summerland.

As the sounds and smells of summer permeate the senses I remember fondly the joys of sharing in adventures with my two young children, exploring and discovering the myriad ways in which Mother Nature shares her most beautiful treasures with us: racing through the forest in the dappled sunlight of a June afternoon, or with our hands in the warm, moist soil of our family garden as we work to bring in the harvest for that evening's meal.

My two sons have always been a great help around the garden, either weeding out (and snacking upon) the lamb's quarters and wood sorrel or gathering the fruits of our labor when they reach that perfect peak stage of ripeness. Working in the gardens has also helped them to develop a diet that's quite different than their fellow classmates at the public school they attend down the road. While the other children's meals are laden with such "luxuries" as macaroni and cheese, potato chips and canned raviolis, my boys have developed quite a taste for freshly harvested broccoli, sugar snap peas still warm from the sunshine and sweet little ground cherries gathered from around the heavily laden vines of the plant.

Children that are given the opportunity to participate in growing their own

SAVING OUR SEEDS: The Practice & Philosophy

food either at home, at school or even at a local community garden are far more likely to sample and include fresh fruits and vegetables in their diets. I found this to be true at home when I watched my then four year old son, Elijah, gleefully snap a head of broccoli clean off the plant and begin munching on the flower buds right there in the garden! As it turns out, kids love green veggies when they have a chance to grow their own!

At our home there was always one row in the garden that seemed to be the exception to this rule, where it didn't matter if our veggies were home grown or not; Elijah never liked to eat tomatoes. He didn't like to eat them fresh, or sliced and sprinkled with salt. He didn't like to eat them on a BLT or hamburger. He didn't like spaghetti sauce and he may have been one of the only kids in America that didn't even like ketchup! No matter how we offered it to him, even if we tried to sneak it into a dish, Elijah just didn't like tomatoes. But, thanks to the amazing diversity of heirloom vegetables, all of that was about to change.

Elijah and I had traveled to the neighboring town of Mt. Pleasant which is home to the Saginaw Chippewa reservation. We were headed to the reservation to deliver corn seed to the 7th Generation building, an indigenous cultural center that is home to a beautiful three sisters garden. If you are not familiar with three sisters gardens, the concept is a simple one. Basically this style of garden is a polyculture of plants that grow harmoniously to the benefit of each. The three sisters are traditionally corn, beans and squash. The corn seed we were delivering on that hot summer day was for the following year's garden plot. It was a special variety called Bear Island Flint, an Ojibwe corn that we acquired from a seed keeping friend in Maine. After a bit of research I found that this variety originally came from an island on Leech Lake in Minnesota. It's a hominy flint corn with twice the protein of conventional corns and it's also high in B vitamins. We were excited to be bringing this gift to our friends on the reservation, but that is a story for another day.

During our visit to the 7th Generation Center, Elijah and I enjoyed a tour of their gardens and other facilities. Just outside the door of the greenhouse, Elijah spotted a very healthy and robust tomato vine climbing its way up a trellis. What really caught his eye was not the deep green foliage of the plant but instead the small, bright yellow, pear shaped fruits hanging in abundant clusters over the entirety of the vines. As curious as always, he hurried right over to this new found discovery and immediately inquired if he could sample the beautiful fruits. I obliged, although I assumed that this tomato tasting experience would be just like any other for young Elijah.

Throughout my many years on this planet, a lesson I am reminded of time and time again is that one should never assume anything; one should always remain open minded. On this day, I was learning that lesson again from my oldest son as he popped that pear shaped tomato into his mouth and I watched as his eyes filled with both surprise and delight! "It's so good!" We quickly filled a small brown paper sack with as many of these little treasures as we could fit and hurried back to the car for the drive home. Elijah snacked on yellow pear tomatoes for most of the half hour we were on the road and then he crinkled the top of the bag shut and set them aside. He informed me that since these new found tomatoes were so delicious, he planned to bring the rest home and harvest their seeds so he could enjoy these tasty fruits for years to come. And that's exactly what he did!

It's been three years since Elijah first tried that yellow pear tomato at the 7th Generation building in Mt. Pleasant, Michigan. He has faithfully grown and tended his tomato plants and harvested their seeds every season since then. It's certainly a proud moment for a parent watching his oldest son grow up as a steward of the garden, maintaining a variety that he has grown to love. But even more than pride, I can't help but feel a nearly child-like sense of wonder at the incredible diversity nature offers that even made this story possible. With so many shapes, sizes, colors and flavors of tomato available, it was only a matter of time before an open minded garden adventurer like Elijah would find a variety that suites his taste buds. Since that fateful summer day, Elijah has sampled and discovered a number of other tomatoes that he enjoys as well.

As I sit here at the kitchen table, staring out my window at the wet, muddy and unseasonable weather we are having this winter in Michigan, I can't help but let my memories take me back to the time when the sun beat down from overhead, the soothing notes of bird songs filled the air and I learned from Elijah how diversity fills our plate.

LIMA

Phaseolus lunatus

Family: Fabaceae

Pollination:
Self-pollinating as well as insect pollinated

Isolation:
200-500 feet

DRY SEED **A ANNUAL**

Photo by Baker Creek Heirloom Seed Company

Potawatomi Lima

This stunning bean has been grown by the Potawatomi Nation of North America since the late 1500's. These seeds have become widely distributed thanks to the efforts of Roger Gustafson, who got them in the 1980's from the Prairie Band of Potawatomi Nation in Kansas.

The flattened, moon-like seeds of lima are much more than the pale green beans we were offered (and likely rejected) as children. The seeds of this beautiful species are white, green, black, red and even speckled and mottled with various combinations of all of these colors. *Phaseolus lunatus* has both bush and pole growth habits with the pole varieties sometimes growing up to ten feet long! These are long season crops as well, so gardeners in northern climates may want to consider starting their plants in advance and transplanting them out to the garden after the soil has warmed to around 65°F (18°C).

Although this member of the Fabaceae family is self-pollinating, its flower structure is very susceptible to crossing via insect visitors. Bagging the blossoms is one option to help avoid this and on bush type plants this can be a quick fix, but with the tall growing pole varieties, you may only be able to bag a small amount that are within easy reach. Each pod will only contain three or four seeds so this option is really just for those that wish to collect a small amount of seed. The best solution to ensure a harvest of true-to-type seed is to only grow one variety per season. Remember though, lima beans will not cross with other species of Phaseolus such as common or runner beans.

Lima beans are harvested and processed as dry seeds. See page 182 for more information on this process. If you decide to shell your limas by hand, consider wearing gloves as the tips of the pods can be sharp.

SAVING OUR SEEDS: The Practice & Philosophy

MELON

Cucumis melo

Family: Cucurbitaceae

Pollination: Insect Pollinated

Isolation: ½ mile

WET SEED

A ANNUAL

Photo by Baker Creek Heirloom Seed Company

MELON

There are many different varieties of melon to choose from for your garden and they are all equally as easy to grow for seed as they are for food. When your melons have reached the market ready stage, they are also ready for seed harvesting. The greatest challenge in melon seed production is simply finding a melon that grows and produces well in your garden.

Cucumis melo is insect pollinated and therefore growing only one variety per season is the simplest method for ensuring the genetic purity of your variety. If you must grow more than one melon variety in a season and you're unable to properly isolate your plants, the flowers will need to be hand pollinated as described on page 172. The process is the same as it is for squash but melon flowers are smaller and more delicate.

Harvesting and cleaning the seeds is a simple process as well, plus you can enjoy eating your tasty melons while you work! Letting the seeds soak in water for a short time will make removing the pulp a bit easier and will also help you to separate the viable seeds from the immature ones. See page 176 for more details on processing wet seeds. It's important to note that some viable seed will still float along with the immature seeds and be lost in decanting, but the large quantity of seeds per melon tends to make up for this. If this is a concern, save all of the seeds to dry and the lighter, under-developed seeds can be removed later with a quick winnowing. Details on winnowing your seed can be found on page 182.

SAVING OUR SEEDS: The Practice & Philosophy

MUSTARD GREENS

Brassica juncea

Family: Brassicaceae

Pollination:
Insect Pollinated, occasionally self-pollinating

Isolation:
½ mile

DRY SEED

A
ANNUAL

Photo by Baker Creek Heirloom Seed Company

MUSTARD GREENS

This spicy flavored green is a very popular crop in the American South but this cool season plant can be grown and enjoyed anywhere. Not only is *Brassica juncea* cultivated for its delicious leaves, the seeds are also grown as a commercial crop, which are then processed into the condiment brown mustard as well as pressed to produce mustard oil.

Mustard greens originated somewhere in central Asia and have been grown in both India and China for well over three thousand years. It's believed that this species developed from an unlikely cross between two other species, *Brassica rapa* and *Brassica nigra*.

Growing mustard for seed is a very easy and straightforward task. The plants are annual, self-compatible and are even occasionally found to be self-pollinating! It's worth noting that in some areas of the United States, *Brassica juncea* has become naturalized and this may pose a risk for growers hoping to maintain varietal purity. If a proper isolation distance from other plants of the species is not possible, the grower can consider caging their mustard and possibly even introducing pollinators to ensure good seed set. Remember, mustard flower stalks can reach up to four feet in height so the isolation cages would need to be built to accommodate this.

Like other members of the Brassicaceae family, mustard will produce its seeds in pods called siliques. When the seeds have matured and are ready to be harvested, the siliques will dry and turn a light brown color. Being careful to avoid shattering the pods during harvest is important, but threshing and cleaning the seeds is a very simple chore. For more information on threshing and winnowing dry seeds, please see page 182.

SAVING OUR SEEDS: The Practice & Philosophy

OKRA

Abelmoschus esculentus

Family: Malvaceae

Pollination:
Self-pollinating and insect pollinated

Isolation:
½ mile

DRY SEED

A ANNUAL

Photo by Baker Creek Heirloom Seed Company

OKRA

While Okra is a crop that many attribute to the southern United States, it can be grown in most places with some success. The trick to a good harvest of fruits is the heat that the plants need to produce and to succeed with a seed crop you need a long season for the capsules to reach maturity. In short season areas, starting your plants early to transplant out when temperatures are warm is a necessity.

There is not much agreement on the origin of *Abelmoschus esculentus*, it would seem likely that it was first domesticated somewhere in Africa. It is documented as a cultivated crop in Egypt sometime in the fourteenth century and made its way to the Americas during the seventeenth century, brought by African slaves that cherished this beautiful and delicious crop.

Okra flowers are incredibly beautiful, resembling the related hibiscus, and although they are self-pollinating, they are very attractive to insects. To avoid cross pollination there are a few options; simply growing one variety per season, isolating your varieties by ½ mile or bagging the individual flowers to ensure seed purity. Each pod will contain quite a few seeds so simply bagging a handful of flowers will still produce a decent amount of seed for next year's garden.

After the okra pods are left on the plant to fully mature and dry, they can be harvested and hand threshed (wear gloves!) or flailed in a sack or bucket before being winnowed clean. More information on processing dry seeds can be found on page 182.

SEEDKEEPER'S TALE

CHRIS SMITH
NORTH CAROLINA, USA
FOUNDER, UTOPIAN SEED PROJECT

Growing up in England, where okra is not well known, I skipped the whole okra childhood trauma thing. I didn't move to America until I was 30 and so also skipped the cultural bias against this maligned Malvaceae. I didn't pre-judge it for its sliminess. I didn't know its history; it was just another new vegetable that I hadn't been able to grow in England. I did come close to falling into the large contingent of clichéd okra haters in 2006. A greasy spoon just outside of Clayton, Georgia was doing no favors for okra's downtrodden reputation. But that was a long-ago vacation and a small blip in my otherwise loving relationship with okra, which truly grew from a single okra pod gifted to me when I was forced to attend my fiancé's bridal shower in Columbia, South Carolina in 2012. The dried okra pod that Belle's friend gave to me had been grown by a family in Rosman, North Carolina for generations. It had a history and, if I chose, a future.

Being gifted seeds is nothing like being gifted a teapot or a silicone spatula. A gift of seeds is a weighty thing. I did not know the full heritage story of those seeds but I have grown and saved them every year since my wedding and at some point started calling it Rosman Wedding okra. Rosman Wedding is a fairly standard green okra, nothing exceptional, although it did surprise

SAVING OUR SEEDS: The Practice and Philosophy

Photo by Baker Creek Heirloom Seed Company

me with superior productivity in a large 76 variety okra trial I ran in 2018, perhaps speaking to Rosman Wedding's local adaptation more than anything else (not a shabby reason in itself to be saving your own seeds).

At the beginning of that trial I was a little concerned that all these okra varieties with their cool names, origins and stories, would turn out to represent a generic phenotype or two. That was not the case. To walk into my okra field was to be astounded by diversity; an astonishment that inspired Sow True Seed to form a spin off non-profit dedicated to investigating and celebrating diversity in food and farming, which we called The Utopian Seed Project. As executive director, I get to throw the standard need for gardening restraint to the wind and say, "I'll take one of everything please!" In 2019, amongst many other exciting crops, I will again grow 70-odd varieties of okra, totally different from 2018, and now I *expect* to see diversity.

Okra is responsible for my passionate seed-to-stem philosophy of food; my champion crop in that regard. The more I read and discover, and grow and experiment, the more I realize just how incredible okra is as a whole plant. The plethora of okra culinary traditions around the world has created a myriad of pod preparations but there are also examples of flower, seed, leaf and stem uses that could easily be incorporated into our own homesteads and farms.

Ethiopia is okra's most likely homeland and many African countries have demonstrated a wide range of uses for the okra plant, from okra leaf soup to okra flour fortified breads. Back in the USA, Clay Oliver Farms in Georgia is pressing an okra seed oil and Smiling Hara Tempeh in North Carolina has successfully cultured an okra seed chickpea tempeh. I like to roast the seeds and grind them to make a rich nutty okra seed flour to use in pancake mixes and pizza crusts. The flower is a lot of fun as a daring edible garnish for almost any dish, but also imparts a vivid red to infusions of vinegar and vodka. I've been experimenting with okra flower bathtub gin with tasty results and delightful colors. Asheville Tea Company included okra flowers into one of their savory tea blends along with roselle (*Hibiscus sabdariffa*) for a mighty Malvaceae combo. Okra pod facials, okra slime hair conditioner, okra cordage, okra paper and dried okra pod decorations are all non-edible projects that my family and I have enjoyed (or tolerated).

As the rabbit hole deepened, I found myself afloat in a vast sea of okra potential. Katrina Blair helped me discover okra pod marshmallows, Chef Steven Goff led me to okra kimchi, and alone I stumbled upon oyster mushrooms grown on dried okra pods (post-seed saving). OWL Bakery in Asheville made an okra flour sourdough and savory okra-corn muffins and Sunburst Chef and Farmer now sell okra microgreens. Truly, this plant is so versatile that I feel the only limitation is one's imagination and a willingness to embrace it.

PARSNIP

Pastinaca sativa

Family: Apiaceae

Pollination:
Insect Pollinated and occasionally self-pollinating

Isolation:
800 feet – ½ mile

DRY SEED **B BIENNIAL**

OVER-WINTERING REQUIRED ...SEE PG 184

Photo by Baker Creek Heirloom Seed Company

PARSNIP

As far as I'm concerned, parsnips just don't get the credit they deserve in today's gardens. I know many people that have never grown them and just as many that have never even included these delicious root crops in their kitchens. It's a shame. Parsnips are versatile, tasty and relatively easy to grow, once you get the seeds to germinate!

Low germination rates aren't the only challenge that gardeners face when growing *Pastinaca sativa* either; the sap exuded from the stems and leaves of the plant can be quite irritating to your skin. For this reason, it is recommended that growers wear long sleeve shirts, gloves and other protective gear when working in the gardens with their parsnips.

Parsnips are biennial and need to be vernalized in order to trigger the plants to flower and produce seed but they are so incredibly cold hardy that they can be left in the ground to overwinter in even the coldest of climates. If your growing region doesn't reach cold enough temperatures to vernalize your plants (under 50°F or 10°C for 10 weeks), then you'll need to dig up the roots and prepare them for storage in a refrigerator. More details on overwintering your crops can be found starting on page 184.

Since parsnips are insect pollinated, they will need to be isolated from other varieties in order to avoid crossing. In some areas, gardeners will also have the challenge of wild parsnips to contend with that will also easily cross pollinate with your garden variety. Since the flowers of *Pastinaca sativa* are perfect and self-compatible, the dedicated seed grower can consider hand-pollinating their parsnips for the production of pure seed. More information on hand pollination is on page 172.

The mature dry seeds can be harvested and cleaned the second season with the same techniques that one would use for carrot seeds. For additional details on processing dry seeds, please see page 182.

PEANUT

Arachis hypogaea

Family: Fabaceae

Pollination:
Self-pollinating and occasionally insect pollinated

Isolation:
50 feet – ½ mile

DRY SEED

A ANNUAL

Photo by Baker Creek Heirloom Seed Company

Argentine White Valencia Peanut

A rare valencia type peanut from the collection of well known seed collector Mr. Blane Bourgeois. This stunning peanut could be the perfect compliment to the Tennessee Red as it has a very similar flavor profile and size.

Peanut products seem to be pretty common items found in most homes; peanut butter, peanut oil, salted and roasted as snacks, but this unique legume isn't quite so easy to find growing in most garden beds… and that's not because these 'nuts' actually grow underground! Peanuts are easy and fun to grow, a gardener just needs around 120 days of good warm weather and a nice, loosely textured soil.

Peanuts were first domesticated in South America, in the region of modern day Peru and Argentina, around 7,500 years ago. Later, the Spanish took peanuts with them to Asia and India and Portuguese traders brought them to Africa. It was then, from Africa, that *Arachis hypogaea* finally made its way to North America.

Peanuts have perfect flowers and are self-pollinating, although occasionally insects will cross pollinate two varieties. There seems to be a bit of disagreement in the severity and likelihood of crossing to occur, with some sources stating that 35% cross pollination is possible for varieties grown within ½ mile of each other, especially between older varieties whose flowers tend to have a more protruded stigma. The easiest solution in this case is simply only growing one variety per season.

Once the perfect, yellow flowers are fertilized, they will drop off and the plant will form a peg which will then grow down into the Earth where the peanuts themselves will develop. The plants will later turn yellow to signal seed maturity and you can simply pull them right out of the ground. The peanuts will now need to cure in a dry location out of direct sunlight for two to three weeks. It's best to leave the peanuts in their shells until you are ready to plant them the next season. If you store your seeds in a refrigerator or other location around 35°F (1.7°C), they will remain viable for three to four years.

SAVING OUR SEEDS: The Practice & Philosophy

Standing Amongst The Corn
By Bevin Cohen

It seems that every year I have a different infatuation in the garden, a different species, or even a variety, that stands out above the rest making my heart flutter with excitement during each stage of its life cycle. After the precious seed germinates and its sprout begins to push free from the soil, my breath quickens. When the first set of true leaves unfurl and reach out towards the sunlight, my palms begin to sweat. As the flower blossoms emerge from their buds and bloom into their full magnificent glory, I tremble with anticipation. And finally, when the fruits of my plants ripen and mature, I can imagine them laden with the seeds of tomorrow's gardens and I can no longer contain myself. The time for seed harvest has arrived!

Over the years my passionate attentions have been claimed by beans, the jewels of every poor man's garden, and Capsicum peppers so diverse and colorful as they glisten in the sun. I've been distracted by the smells and flavors of heirloom tomatoes, still warm from the afternoon heat. Last year my love affair was with kale and collards and this season my oldest son Elijah has felt Cupid's arrow pierce through his heart in the name of Okra; he's trialing at least a half dozen different varieties in his garden.

But despite the waxing and waning of my garden passions and my tendency to throw myself into committed vegetable relationships with whatever 'flavor of the week" suites my fancy, there's one garden crop that has continually stood by my side, a silent partner always waiting for me to finally settle down and realize everything that she has to offer. She has never asked for anything more from me than just simply my attention, and maybe some water time and again. This steadfast and reliable beauty is none other than corn, the amazing *Zea mays*.

SAVING OUR SEEDS: The Practice & Philosophy

On our small farm in Michigan we have grown quite a bit of sweet corn, as well as flint and flour types. More recently, the boys and I have started a small scale popcorn breeding project, which is really just a fun opportunity for everyone to learn a little bit more about plant genetics. Being a wind pollinated plant, we simply alternated plantings of a few choice varieties of popcorn to allow them to cross pollinate with each other to give us some interesting genetics to select from. It's an entertaining and educational experience and when fall arrives, we get to enjoy some home grown popcorn. It's a win-win for everyone.

One of my favorite activities during peak season is to walk the many rows of the gardens at sunrise, to enjoy the silence of the morning before the birds signal the start of the workday with their cheerful whistles and songs. This part of the day leading into the rising of the morning sun is filled with quiet potential and I find it to be the best time to not only collect my thoughts, but also to be still and listen to the world as it awakens.

It is often here in the garden, as I'm standing amongst the corn, that the first winds of the day take flight and blow their way through the upright and proud rows of my sacred garden companion. As she rustles in the breeze, it is almost as if I can hear her voice, whispering in my ears, singing me the songs of a thousand generations that stood before her. It is through this corn, the most sacred of all plants, that people have been sustained on this land, through her nourishment and care. To hear the songs of the corn carried on the morning breeze, I am reminded of the hard work and sacrifice of the people that walked this land long before I ever planted a seed in the Earth's womb. From one generation to the next, from the hand of the grandmother to the hand of the granddaughter, the seeds of life have been passed and as I stand here in the field, I know that I too am tasked with this most serious of responsibilities. To plant seeds in the ground is to make a commitment to life itself, and as I nurture this plant and watch it grow, I agree to adhere to the covenant of Grandmother Corn. I have now joined the lineage of the seed keeper. If I am to plant the seeds, I must also agree to save the seeds. There is no difference between the two and the circle of life continues.

PEA

Pisum sativum

Family: Fabaceae

Pollination:
Self-pollinating and occasionally insect pollinated

Isolation:
10-20 feet

DRY SEED

A ANNUAL

Photo by Baker Creek Heirloom Seed Company

Golden Sweet Pea

One of the few yellow edible podded peas available. This rare garden delight was first collected at a market in India and has continued to stun gardeners around the world with its bi-color pink and purple flowers and heavy production.

Our kids absolutely love growing snap peas every spring. They pick so many of them to snack on in the gardens that it's rare to see any peas make their way in to the house at all! I certainly don't blame them one bit, fresh garden peas are one of my favorite spring treats too.

Pea varieties are roughly divided into a number of categories, the first one being field peas (grown for fully mature dry seeds) and garden peas (harvested immature to be eaten green). Garden peas themselves are also further divided into different types; those with inedible pods (shelling peas) and those with edible pods (snap and snow peas). All of these types of peas are handled in the same manner when harvested as seed crops, regardless of the growth habit or the variety's market maturity. It's interesting to note that the term field pea is also commonly used to describe cowpeas, which are a different species, *Vigna unguiculata*. More information on cowpeas can be found on page 54.

Pisum sativum plants meant for seed are simply left to mature in the garden until the pods are fully formed. They will then turn brown as they dry. If the weather is calling for rain and you are concerned about your seeds getting wet in this crucial drying stage, you can pull the plants, or harvest the pods individually if working with a small quantity, and hang them somewhere out of the elements to finish drying down. After the peas are fully dry, they can be threshed and winnowed like any other dry seed as described in detail on page 182.

PEPPER

Capsicum spp.
C annuum, C chinense,
C baccatum, C pubescens,
C frutescens

Family: Solanaceae

Pollination:
Self-pollinating and insect pollinated

Isolation:
1,500 feet to ½ mile

WET SEED **A ANNUAL**

Photo by Baker Creek Heirloom Seed Company

The sheer diversity of the Capsicum genus is astounding! The five most common cultivated species of peppers found in most gardens offer an incredibly wide range of colors, shapes and sizes. They also offer flavors from sweet and tangy to burn your tongue off spicy! With this much diversity, any gardener can find a variety or two of Capsicum that is just right for them.

It's important to note that, with peppers, inter-species cross pollination can still occur, sometimes at a rate of up to 80%. The only exception to this is the *C. pubescens* species, noted by its black seeds, as they are not compatible to crossing with other members of the genus. Also, although the perfect flowers on pepper plants are self-pollinating, they are very attractive to insects that will happily cross pollinate your varieties as they enjoy their day buzzing around from flower to flower. While most growers don't have the space required to properly isolate their plants, pepper blossoms can easily be bagged to ensure seed purity. In fact, there have been times that I have bagged entire plants with 'lettuce bags' that are available for purchase through a number of sources. I have found this to be the easiest way to avoid crossing amongst my pepper varieties. Row cover also works very well for this.

Harvesting the seed from your fully ripened, mature fruit is quite simple and your technique can be determined by the variety you have grown. For peppers with thin walls, like cayenne types, you can string the pods together and hang them to dry, similar to how you would prepare them to make spice. Once the fruits are fully dried, they can be broken or crumbled to release the seeds. Varieties with thicker walls cannot be dried in this way, but can be cut open and the seeds scraped out onto fine mesh screens or even paper plates and left to finish drying. Very small peppers can be run through a food processor, with a small amount of water, using a dough blade attachment to release the seeds. Once processed, simply decant the pulp until only seeds remain, strain and place out to fully dry.

Remember: Be sure to wear gloves when harvesting seeds from hot peppers as the capsaicin can be very irritating to the skin.

POTATO

Solanum tuberosum

Family: Solanaceae

Pollination:
Self-pollinating and insect pollinated

Isolation:
500 feet

DRY SEED

A ANNUAL

Photo by Baker Creek Heirloom Seed Company

POTATO

Potatoes are an important staple crop around the world and were first domesticated in the mountains of Peru more than 10,000 years ago. Peru is still home to an astounding amount of potato diversity, somewhere around three thousand different varieties! While most of the potatoes grown commercially and in home gardens are grown through vegetative propagation, planting pieces of actual potato and therefore producing clones each season, this entry refers to the important and very interesting work of growing potatoes by true seed.

Many *Solanum tuberosum* plants will produce small berries, similar in appearance to unripe cherry tomatoes. When mature, some of these berries will remain green while others will turn shades of brown or even purple to signal their maturity. You will know your potato berries are ripe when they become softer to the touch and they will even occasionally drop from the plant when they are ready to harvest. You can collect your berries and leave them to sit on a counter or windowsill in your home to further ripen if you'd like. I've left berries in a basket for weeks until I had the time to extract their seeds and they held up just fine. When they are ready, the easiest way to harvest the seeds is to run the fruits, with a little bit of water, through a food processor using a dough blade. The viable seeds will sink to the bottom and you can decant off the berry pulp. It is recommended at this point to ferment the potato seeds similar to how you would process tomato seeds, see page 176 for more details on this.

The diseases that are found in some potato populations are carried from one generation to the next via the tubers; avoiding this is one of the main benefits to growing your potatoes from true seed! Another interesting detail about growing from true seeds is that the offspring will not grow true to type, unlike most members of the Solanaceae family. What this means is that each seed will produce tubers that are in some way different than the mother plant; a wonderful opportunity for you to experiment with plant breeding and to possibly even discover a new variety with superior performance in your gardens! The first harvest grown from true potato seeds will be mostly small tubers which can then be selected from and the best specimens grown the following year to produce a regular crop of your new variety!

SAVING OUR SEEDS: The Practice & Philosophy

SEEDKEEPER'S TALE

CURZIO CARAVATI
WISCONSIN, USA
FOUNDER, KENOSHA POTATO PROJECT

"The Botany of Desire" was published by Michael Pollan in 2001 and a few years later, in the chapter about potatoes, I discovered that the author had a plant growing in his garden which produced berries. In my early gardening years, someone suggested I remove the flowers from potato vines, so to provide more energy to the plants' tuber production. It turns out to be an old wives' tale, not true! Enjoy the diversity and beauty of potato flowers and eventually you may find a fruit berry hanging on.

I didn't get any berries because my potato varieties were not fertile. I didn't own the correct varieties. But eventually I found fertile varieties, as I started to build my potato collection. Seed Savers Exchange Members Bill Minkey and Will Bonsall provided the initial collection stock, then AgriCanada and the International Genebank in Sturgeon Bay, Wisconsin helped me fine-tune my collection. Today I have hundreds of different potato plants which produce berries, also known as seed balls, the most poisonous part of the plant. The berries look like green cherry tomatoes, and you surely want to educate your children, they are not edible!

SAVING OUR SEEDS: The Practice and Philosophy

Photo by Sam Jones

Facebook provided the perfect platform to reach thousands of potato enthusiasts all over the world. Potato is grown everywhere, and therefore the Kenosha Potato Project Facebook Group may be able to complete its global following by eventually acquiring members from Antarctica. Certainly, the mission is well underway in my dream to create a legacy as Johnny Potato Seed! Like Johnny Appleseed, who distributed apple seeds to promote the planting of fruit trees in America, Kenosha Potato Project has several ongoing activities to get gardeners and farmers to try their hands at growing potato, starting with botanical seed, which is also known as TPS (True Potato Seed). In 2019, we were able to distribute the same collection of seed to over 60 growers in different climate zones.

Starting a potato plant with a seed, like you do for tomatoes, is a very interesting learning experience. First, you'll realize TPS potato seedlings are very delicate until it's time to transplant them into the field. The greatest surprise comes at harvest time: each plant is genetically different, and therefore the tubers are likely different. Imagine having 60 growers with about 100 seeds each, like I mentioned above. At harvest time we would have 6,000 new potato varieties! Well, not actually. Many of those plants will not survive the growing season. Some may not even produce any tubers!

Each plant's tubers are, in fact, a different strain which will have to be grown out for several years in order to be evaluated; consistency of tubers' shape and colors, flesh texture and culinary features, storage strength – perhaps the most important feature, as who cares to have a crop which is hard to store? There you go. You are now a hobby potato breeder! The selection of the best strains will eventually lead to your own varieties! Locally adapted, fertile and setting seed berries, preferably every year. Now you have a real family heirloom. Imagine the seed you extract from a potato berry in the year 2020, this seed will germinate better in 2022 than in 2021. It's rare to find seed which improves germination as it gets older. How many seed

types do you know which have the potential of germinating after 70 years? Properly stored seed may be germinated by your great grand-children. Isn't this a wonderful legacy to leave?

On top of that, for instance, the propagation of potato starting with TPS every five years is a very sustainable growing method. Please allow me to introduce an alternative meaning to the word sustainable: Growing potato with tuber pieces is not long term sustainable because tubers are subject to viral infections, which over the planting seasons reduce the yield, eventually down to zero.

Botanical seed (TPS) cannot be infected by common potato viruses and therefore following a five year renewal of the tuber seed stock, any grower may be able to start over with guaranteed healthy seed tubers. This is sustainable as you are no longer dependent on seed tuber producers. The Kenosha Potato Project is just one of the several horticultural and beekeeping projects which are curated at KUFI - Kenosha Urban Farming Institute.

I hope your future potato crops are rich and plentiful!

RADISH

The delightful radish has been grown and enjoyed by gardeners around the world for thousands of years. The most well know varieties grown today can be broken down into two basic groups, the summer radishes which are typically annuals and the winter radishes that are mostly biennial crops. The summer types are the quintessential round, red globes that we are all familiar with but they also come in a number of other colors and shapes; pink, white, purple, bicolored and oblong, tapered, flattened and oval. There is also a third grouping of radishes grown specifically for their larger edible seed pods, the most well-known of which is a variety named 'Rat's Tail'. It's important to note that the seed pods of all radishes are edible, and delicious, when they are young and fresh before the pod walls thicken and become more fibrous.

Although *Raphanus sativus* have perfect flowers, they are self-incompatible and therefore need to be pollinated by insects. To ensure good seed set, it's helpful to maintain a decent sized population of plants left to flower, preferably around at least twenty plants. It is difficult to avoid cross pollination between varieties, so only allowing one variety to go to flower per season is the best method to avoid crossing.

Some radish plants will grow as tall as four feet as they flower and produce seed. The seeds will form in siliques that look very similar to the seed pods of other members of the Brassicaceae family like kale and cabbage, but have much thicker walls and are less likely to shatter before harvesting. Because of these thicker walls, threshing may take a bit more effort, but winnowing the heavy seeds of radish is a simple task. For more information on threshing and winnowing dry seeds, see page 182.

RUNNER BEAN

Phaseolus coccineus

Family: Fabaceae

Pollination:
Insect Pollinated and occasionally self-pollinating

Isolation:
½ mile

DRY SEED

A ANNUAL

Photo by Baker Creek Heirloom Seed Company

RUNNER BEAN

I'd think that everyone can agree that runner beans are one of the most beautiful bean plants growing in the garden and if you haven't grown *Phaseolus coccineus* before, it is time to start! The most popular variety of this species is most likely 'Scarlet Runner', which gets its name from the color of the incredibly showy flowers that the plants will produce, but there are a number of great runner beans out there that flower in shades of white, pink, red and even two-tone. These flowers are very attractive to insects and also to hummingbirds, which to me is reason enough to grow these fast growing pole type beans.

Not only are these plants prolific and pretty but they also produce a significant amount of food. The seeds from runner beans are quite large and can be eaten as fresh shelly beans or dried to be enjoyed later in the winter. The immature pods are also edible when they are still quite young and it is said that even the leaves of this plant are edible! The tuberous root, on the other hand, is not for consumption but can be dug up and over wintered to give you a head start on your runner bean crops the following season!

Runner beans are most commonly pollinated by insects and yields are significantly smaller when blossoms are bagged to avoid insect interaction. For this reason, it is best to isolate your varieties or to consider only growing one runner bean per season if you'd like to avoid cross pollination. These beans, when fully matured and dry, can be threshed and winnowed like any other bean seed crop but take some care as to not shatter the large seeds during processing. More information on harvesting and cleaning dry seed can be found on page 182.

SOY BEAN

Glycine max

Family: Fabaceae

Pollination: Self-pollinating

Isolation: 10-20 feet

DRY SEED

A ANNUAL

Photo by Baker Creek Heirloom Seed Company

Midori Giant Soy Bean

One of the largest, earliest and heaviest producing soybeans available. Perfect for use as edamame and an ideal crop for both large scale and small scale agriculture.

My family and I currently live in the Midwest where million acre, monoculture soybean fields are the norm. Because of this, in my mind, crops like *Glycine max* have become synonymous with many of the problems in our food system brought upon us by this modern, large scale agricultural model. Considering just how much land is dedicated to the production of soybeans, it's hard to believe that this crop only first made its way into North America as a commercial food product in the late 1800s. In such a short time, it has dominated the landscape.

While most large scale farming of soybeans is for animal feed and oil production, I have found a number of home gardeners growing varieties that are meant for fresh eating in the green stage, sometimes known as edamame. Whether you choose to grow *Glycine max* for fresh eating or for dry seed use, it's very easy to maintain varietal purity as soybean flowers are self-pollinating and crossing due to insect involvement is very rare. Let your plants grow to full maturity, waiting until the beans have fully dried and turned brown, and then harvest, thresh and winnow as you would any other bean. More detailed information on threshing and winnowing dry seeds can be found on page 182.

The pods of more modern varieties are quite shatter resistant, but older types may shatter and drop seed in the field, so harvesting before that happens is crucial. If you plan to shell your soybeans by hand, be sure to wear gloves as the dried pods can be quite sharp and rough. Soybeans also require a more gentle touch than other bean species as the seed coats are quite thin and easily cracked.

SAVING OUR SEEDS: The Practice & Philosophy

Potential in an Empty Field
By Bevin Cohen

• •

A few months back, I was driving along some back roads on my way to an event where I was scheduled to speak. It may have been the grand opening of a new seed library or something of that nature but to be perfectly honest; after all of the hours I spend on the road, sometimes the memories begin to blur together. Where I was headed that day is not important, what matters is what I could see out of my window as I rolled along down the road.

Traveling by car through the Midwest can be a terribly boring experience as a great majority of the countryside is dedicated to the large-scale production of commodity crops, namely corn and soybeans. If you are lucky enough to be headed through during peak summer or even during the early fall, you'll be treated to an endless sea of green, sprawling for millions of acres to the horizon in every direction. It may not be the most interesting thing, but at least it's green. This particular trip that I speak of had taken place in the early spring, while these large commodity crop farm fields still sat barren. Bare, not a blade of grass or anything to be seen, nothing but light brown and dusty Earth for miles.

When faced with endless miles of dry empty stretches of land in every direction, only broken up occasionally by small rural towns that consist of nothing more than a gas station, a church and maybe a dollar store or bar, one's mind can't help but wander to thoughts about what purpose these fields are meant to fulfill and I heard myself asking the question, "Why? Why aren't these fields being utilized to their greatest potential?"

To answer this difficult question, we need to understand what these fields are actually being used for right now. Like I mentioned above, a quick drive through this farm country in the summer will reveal that most of this land has been dedicated to the production of corn and soybeans, but ultimately how are these crops being used?

First, let's talk briefly about soybeans. It's estimated by the USDA that within the next few years soybean production will surpass corn and become the number one acreage crop here in the United States. While approximately half of the soybeans harvested are meant for export to China, the EU, Japan and other countries, roughly 70% of all the soybeans grown in the states are used for animal feed. So, in reality, these soybeans do eventually make it to the dinner table, just in more of a roundabout way. The oil that is extracted from these seeds also has a multitude of uses including biodiesel, lubricants, crayons and candles and it can also be found at the grocery store in baked goods, processed foods, in margarine and also marketed as vegetable oil. In 2018 farmers planted a total of nearly 90 million acres of soybeans in the United States.

In that same year, farmers also grew approximately 90 million acres of corn and the greatest use of this harvest was again for animal feed, followed closely by ethanol fuel production. About 10-12% of the corn grown in the United States is used in the production of high fructose corn syrup, sweeteners, starch and cereals. The sweet corn that many of us are familiar with as a delicious summer treat accounts for about 1% of all corn grown in the U.S. annually. Although we certainly do export corn to other countries, it is a much smaller amount than that of soy at an approximate 13% of total harvest annually; most of the corn grown in the states stays in the states.

Unlike soybeans, that were first domesticated in Asia, corn originated in North America and was first domesticated for use in agriculture in the area now known as Mexico more than eight thousand years ago. The indigenous people of this land have maintained a long and sacred relationship with this

special plant. Corn is an important part of many creation stories and it was truly the food that sustained the people, as the people themselves sustained the plant. When European immigrants came to live on this continent, they also adopted corn as a staple crop and found it to be quickly adapting and suitable for growing in various climates and soil conditions.

Over time, corn has experienced what some folks may consider 'improvements' through hybridization efforts as well as genetic modification. The purpose of this book is not to take a position on whether or not this is right or wrong, I'll leave that to the reader to determine for themselves, but I do want to take a moment to share some thoughts on the wide spread commercialization of corn.

In a way, what has happened to corn in the United States also makes me think about tobacco, another plant held sacred to this continent's indigenous population. Some Nicotiana (tobacco) cultivation sites have been found in the area of Mexico that date back to around 1400 BCE and it was quite popular with many native tribes across the land. It was, and sometimes still is, used as an item for trade as well as for use both socially and ceremoniously. In many Indigenous traditions tobacco is seen as a gift from the Creator and is utilized in various prayer ceremonies.

So, to summarize, both corn and tobacco are native to North America and have a very long history of use with the native people that live here. Both of these plants are considered to be sacred and are used in daily life as well as in a spiritual context. After the arrival of the European immigrants, these two revered plants quickly became commercialized cash crops, suddenly considered by these immigrants to be nothing more than mere commodities to be bought and sold. As I drive by thousands of acres of commercial, cash crop, 'bushel per acre' commodity corn, each of those plants reminds me of the mistakes we have made on this land: the atrocities forced upon this continent's native people and the mistakes we continue to make each season as we devote millions of acres of our precious land to produce nothing more than feed for cattle, fuel for our automobiles and oils and syrups for our processed food consumption. Every acre I pass is a stark reminder of a food system gone awry.

While I offer no specific solution, it is imperative that we acknowledge these realities. Perhaps, even by understanding and admitting to the errors of the past, we can shift our perspectives and began the journey towards repairing and healing our relationship with the Earth and these ancient crops that have sustained us for millennia.

SPINACH

Spinacia oleracea

Family: Amaranthaceae

Pollination:
Wind Pollinated

Isolation:
1 - 2 miles

DRY SEED

A ANNUAL

Photo by Baker Creek Heirloom Seed Company

Bloomsdale Long Standing Spinach

The Landreth family introduced this variety in 1826, and it became an instant success, now a classic. Known for its outstanding ability to withstand temperature changes, it won the prestigious AAS award in 1937.

Spinach is the cool season crop that always causes the most joy and cheering from our children every spring, they absolutely love this delicious vegetable! They don't want it cooked, or from a can like the cartoon tough-guy Popeye, they just want it fresh… and they want it by the handful.

Spinacia oleracea is simple to grow and, with a little bit of understanding, it's also easy to harvest as a seed crop. The first challenge to overcome when growing spinach for seed is the impulse to pull the plants when they begin to bolt as warmer weather arrives. It's also important to maintain a good sized population of plants in order to ensure good seed set because spinach plants are dioecious, which means there are separate male and female flowering plants. Being wind pollinated, the likelihood of cross pollination with other spinach varieties is high so it is always easiest to only grow one variety per season to ensure varietal purity. Since most people tend to pull their spinach plants once they begin to flower, it is unlikely that your neighbors will be growing their spinach plants to seed, but it is always a good idea to find out. If cross pollination with a neighbor's spinach is a concern, consider using a spun-poly row cover to isolate your plants, the pollen being too small for mesh screens or baggies to be effective.

Once dried, the seed-laden stems of the female plants can be cleaned by hand or flailed in a bucket and then winnowed to clean away the plant debris. More information about processing dry seeds can be found on page 182. If you are harvesting your spinach seeds by hand, and you are growing a variety with prickly seeds, be sure to wear gloves while working.

SAVING OUR SEEDS: The Practice & Philosophy

SQUASH

There are five domesticated species of squash that are commonly grown in home and market gardens. All five are included in this singular entry because although there are certainly differences from one to the next, the methods for growing, harvesting and processing the seeds remains the same.

All five of these species were first domesticated either in the region of Mexico or further south into South America. They are all monoecious, having both male and female flowers on each plant, are self-compatible and insect pollinated. Inter-species pollination is not a concern, although different varieties within the same species will cross with one another. It's important to know what species you are growing each season to avoid cross pollination from happening. Hand pollination is certainly an option for people that are growing multiple varieties of the same species or that garden in areas where isolation from other growers is not possible. Hand pollination techniques are described in more depth on page 172.

Squash seeds can be harvested from the fruits when they have fully matured. After collecting mature squash from the gardens, leave them to age a bit longer, around four weeks, to allow time for the seeds to finish developing. The seeds will then be collected and cleaned, then dried and put away until next year. The process of harvesting wet seeds is described in detail on page 176.

See next page for more detailed information on each squash species.

Curcurbita pepo: This is likely the most common squash species grown in the garden and was first domesticated in Mexico around 10,000 years ago. All summer squashes are likely to be *C. pepo* as well as a few winter types. It's important to remember when planning to harvest seed from summer squashes that those particular fruits must be left to mature and ripen well past what we consider market ready.

Curcurbita maxima: What I would consider to be the second most popular garden squash, *C. maxima* was first domesticated in South America and was cultivated in the area now known as Peru around 2000 BCE. This species has also become quite popular in Australia where it has been grown since 1788 and includes a number of well-known varieties including 'Hubbard', 'Buttercup', 'Banana' and 'Kabocha'.

Curcurbita moschata: It is believed that this species was domesticated and first cultivated in the area around southern Mexico about 4900 BCE. This species includes the popular butternut style squash and is probably my favorite species of squash to grow. The tasty and rich dark orange flesh is satisfying and nutritious.

Curcurbita argyrosperma: This species of squash is also believed to have been first domesticated in southern Mexico, around 3100 BCE. The varieties of *C. argyrosperma* that are most well known in the gardens are cushaw types as well as silver-seeded gourds. The opening ceremony at the annual Appalachian Seed Swap in Kentucky includes the 'cutting of the cushaw' and this special moment was actually my first introduction to this species of squash.

Cucurbita ficafolia: This squash species is sometimes referred to as fig-leafed gourd, due to the shape of the plant's leaf. Likely domesticated in the area known as Peru, this squash is enjoyed around the world for its edible leaves, seeds and fruits. Some varieties of *C. ficafolia* even contain black seeds.

SQUASH

Photo by Baker Creek Heirloom Seed Company

SEEDKEEPER'S TALE

SARAH TOMAC
MICHIGAN, USA
OWNER, TOMAC PUMPKINS

When you visit my farm in the fall, it's like an attack on the senses. There is color everywhere. I am the fourth generation farmer on a cash crop farm that just happens to specialize in 200 some odd varieties of pumpkins, squash and gourds. The first time visitor is awestruck by the color, size and types of pumpkins everywhere.

Of course the first comment we hear is, "I didn't know there were this many different types" and the second is, "How did you get into this?"

There are over a thousand different types of pumpkins and more all the time being cultivated thanks to crossbreeding and genetic mutations and local adaptations. It's a wonderful world I live in that I get to play with crossing pumpkins and keeping strains pure.

How did I exactly get started with these pumpkins? I'll start with, it is in my destiny. There is a photo of me at two years old, sitting amongst a pile of pumpkins, eating a cookie (my second favorite thing to do) while my mom helped my older brother get his exhibits ready for the county fair. Our county fair has a pumpkin for its logo and a long time ago, it was *the* fair to bring your giant pumpkin to for bragging rights. That was the thing. Giant

SAVING OUR SEEDS: The Practice and Philosophy

Photo by Sarah Tomac, Tomac Pumpkins

pumpkins. And it became my thing, pumpkins. When I became old enough to exhibit at the fair, pumpkins were the only thing I wanted to take. It didn't matter to me about the livestock project, sewing or the crafts or the baking or even the other crops that I hauled to the fair every year. What mattered was the week I got to skip school to take my pumpkin to the fair. As I got older, I learned to grow more so I would have a better selection to compete with against my brothers. As the only sister, I had to win, not my brothers.

What do you do with a bunch of extra pumpkins when you're ten years old? You ask grandma and grandpa really sweet like if you can sell the extras on the side of the road at their place of course! My grandparents said yes because, well, who can deny a kid who wants to earn some pocket change?? It wasn't anything giant to begin with. Just a few orange field pumpkins on a hay wagon on the roadside. Until one year, when we were baling straw and a car stopped in and asked the price to purchase a bale. After car number three stopped that day, my parents decided that they should set up a more permanent straw sales spot in the yard. Later in the fall, with the pumpkin wagon and the straw wagon next to each other, someone stopped in and looked at the corn field growing right next to the farm buildings and asked if they could have a bundle of corn stalks to decorate their yard, along with the straw bale and pumpkins. Well, now we sell corn stalk bundles, straw and pumpkins.

The diversity of pumpkins available for purchase didn't start right away. After a few years, we added butternut and acorn squash, because people were asking for them. We had a couple of different types of common squash on offer and were just enjoying the pocket money.

It wasn't until my surprise baby brother was born and he received the board book, "Paddy's Pumpkin Patch". In the book, Paddy cooked with nothing but pumpkin. Pumpkin burger, pumpkin fries, pumpkin cake, pumpkin ice cream, pumpkin shake, and roasted pumpkin. I started wondering about how

SAVING OUR SEEDS: The Practice and Philosophy

SEED KEEPER

many different ways that one could eat pumpkin. About the same time, I had the opportunity to travel overseas to Australia. While over there, I learned just how delicious a different pumpkin was. I sent some seeds back to mom with the instructions: 'plant these and eat'. She was doubtful at first, then soon learned that it was a really yummy pumpkin, even though it was blue/gray on the outside. We found out by accident one year that it was also a great keeper …but that's another story.

The extra pumpkins were added to the types for sale and that began the journey. Thirty years ago, even twenty years ago, or even ten, to order different types of pumpkins and squash from a seed catalog; there wasn't a whole lot to choose from. We had to source things in a much different manner. We had to save our own seeds.

The expansive variety of pumpkins we offer has really has taken off in the last few years since I've moved home. Maybe it's my need for something different and new in the pumpkin world every year or maybe it's because my little brother is friends with someone who travels the world for a living collecting seeds. Maybe it's because we pioneered stacking pumpkins long before anyone else thought about it. Neighbors always said we were ahead of the trends anyhow, might as well stick with that.

Any way that you look at it, I have to hand pollinate flowers in order to keep those pure strains alive and well. I have some pretty rare and hard to find seeds on the farm. Planting one or two pumpkin seeds from some unknown place and not knowing anything about them gets pretty exciting at times. All summer, I will wait impatiently for the first leaf, the flowers, the first fruits and the harvest. A pumpkin is not just a pumpkin; it's the world's oldest form of superfood. Pumpkins provide the consumer with a major portion of their daily nutritional needs. They are full of omegas and cancer-fighting, anti-aging, immune boosting goodness.

My visit to a farmers market is not complete without the question from a future pumpkin lover, "Which one is your favorite?" I always need more information before I can provide a worthwhile answer. I can't answer with just one type of pumpkin when there are so many different varieties to choose from! Do I want one that is savory, sweet, tender, dry, flaky? How will it be cooked? Will the pumpkin be grilled, roasted, fried or sautéed? Is it for soup, Sunday roast, grilled with burgers on a tailgate, salad, rice, pasta, chicken, lamb, pork or kangaroo? What are we drinking with it; water, tea, milk, wine, beer or juice? Is this for breakfast, lunch, dinner or snack? I sound like a wine taster or microbrew judge with the number of questions that I have!

In reality, I have a few that I really like and consistently eat. Those are my personal favorites and every year, I seem to find another one to add to the list. Right now I'm up to the following depending on my meal: Bliss, Kent, Moranga, Queensland Blue, Greet Sweet Red, Honeynut, Black Futsu, and Camillo (for the seeds!!!) And, of course, there are more. It just depends on my meal!

Photo by Sarah Tomak, Tomak Pumpkins

SUNFLOWER

The beautiful annual sunflower is a majestic and multipurpose addition to any garden space. With varieties growing anywhere from three to six feet tall and offering a range of flower colors including cream and light yellow to fiery red and even purple, there is surely a sunflower to fit every gardener's tastes.

Not only visually appealing, the sunflower also boasts a multitude of uses. *Helianthus annuus* is obviously a decorative addition to your growing area but there are varieties grown specifically for their edible seeds and others types known as oilseed crops that are used for birdseed and sunflower oil production. In fact, when young and immature, even the leaves of the sunflower plant are edible and delicious!

It's interesting to note that the sunflower is not a single flower, but actually a cluster of smaller flowers arranged together on the head. Although the flowers themselves are perfect, they require insect activity in order to be pollinated. For this reason, it is important to keep your sunflower variety isolated to ensure production of pure seed that will grow true-to-type. Older varieties tend to be self-incompatible and therefore a grower will need to have at least 20-50 plants to ensure they are preserving genetic diversity and getting good seed set.

Sunflowers can be hand-pollinated if needed, but one will need to be sure to keep the flowers bagged for the entirety of the flowering cycle, usually up to 10 days. Hand pollination can be accomplished with a paint brush using techniques as described on page 172, or the method can be as simple as removing the bags and rubbing neighboring flowers together briefly.

When sunflowers have matured and the seeds are ready to harvest, the back of the flower head will turn from a green color to a yellowish brown. Try to leave the flowers out to dry as long as possible. If rain or birds threaten your harvest, cut the flowers and move them to a covered location to finish drying. Once dried, they can be removed from the heads simply by rubbing them by hand or using hardware cloth to dislodge the seeds. If needed, the sunflowers seeds can then be threshed and winnowed before being stored in a cool dry location until next season.

SAVING OUR SEEDS: The Practice & Philosophy

TEPARY BEAN

Phaseolus acutifolius

Family: Fabaceae

Pollination: Self-pollinating, insect pollinated on rare occasion

Isolation: 20-50 feet

DRY SEED

A ANNUAL

Photo by Baker Creek Heirloom Seed Company

TEPARY BEAN

Domesticated somewhere in the Sonoran Desert of northwest Mexico over 5,000 years ago, the tepary bean is the short season, heat tolerant solution to growing beans in areas experiencing increased drought pressures. In many ways *Phaseolus acutifolius* are very similar to other beans of the same genus; they are grown for their mature seeds, they can have growth habits of either bush types or vining, and they also have perfect, self-pollinating flowers that are very unlikely to cross with other varieties grown in the nearby vicinity. In fact, tepary beans are one of the most consistently self-pollinating crops.

What sets this particular species apart from its brethren is its love for hot, dry climates. Even when exposed to extreme drought like conditions, tepary beans are still able to produce a good harvest which can be eaten as immature green pods, as shelly beans or, most commonly, as dried beans stored for later use.

Another benefit of growing tepary beans is the relatively short season needed to mature the dried seeds. Many varieties are known to flower within 30-40 days after planting and can produce a harvest within 70-90 days.

Like other beans grown for seed, tepary should be harvested and processed as a dry seed, after the pods have dried down and begun to turn brown. Much like their wild relatives, many varieties of the species still retain the tendency of the pods to shatter at maturity; care must be taken at harvest time to ensure the seeds are not lost. Because of this tendency to shatter, pods can be harvested a little earlier and shelling them is incredibly easy to do by hand. If you have grown a large quantity that needs to be processed, threshing and winnowing the seeds is fairly simple. You can learn more about processing dry seed on page 182.

TOMATILLO

Physalis philadelphica

Family: Solanaceae

Pollination:
Insect Pollinated

Isolation:
½ mile

WET SEED | ANNUAL

Photo by Baker Creek Heirloom Seed Company

Purple Coban Tomatillo

A strikingly gorgeous heirloom variety collected in the beautiful mountain town of Coban, Guatemala. While the degree of purple coloring may vary from one fruit to the next, the sweet and distinctive flavor profile is consistent throughout the harvest.

Although a different genus than tomatoes, tomatillos are in the same family and are cultivated in a similar manner as their plant cousin. In fact, the common name tomatillo actually translates to "little tomato". It is believed that the fruits of this sprawling plant were enjoyed by both the Mayans and the Aztecs and *Physalis philadelphica* has enjoyed a long history of culinary use throughout Mexico and Guatemala.

Like other members of the genus Physalis, the fruits of tomatillos are encased in a paper-like husk that will eventually be split open by the delicious ripe fruit when mature. These ripe fruits are commonly green but are also sometimes yellow or purple or even, in rare occasions, red.

Tomatillos are self-incompatible and therefore the gardener must grow a multitude of plants to achieve good pollination and ensure fruit set. It is recommended that one grows at least six to ten plants in a season to also maintain the genetic diversity of the population. Being insect pollinated, isolation is needed in order to avoid crossing with other varieties. While bagging and hand pollinating is certainly possible, simply growing only one variety each year is the easiest method to avoid cross pollination. For growers interested in learning more about hand pollinating their tomatillos, more information can be found on page 172.

When the tomatillo's fruits are mature and ready to be harvested for the kitchen is also the time for collecting their seeds. Tomatillo seeds are processed like other wet seeds from small fruits as discussed on page 176. The easiest technique uses a blender to remove the small seeds from the juicy pulp of the fruit which are then put out on small screens or paper plates in order to dry.

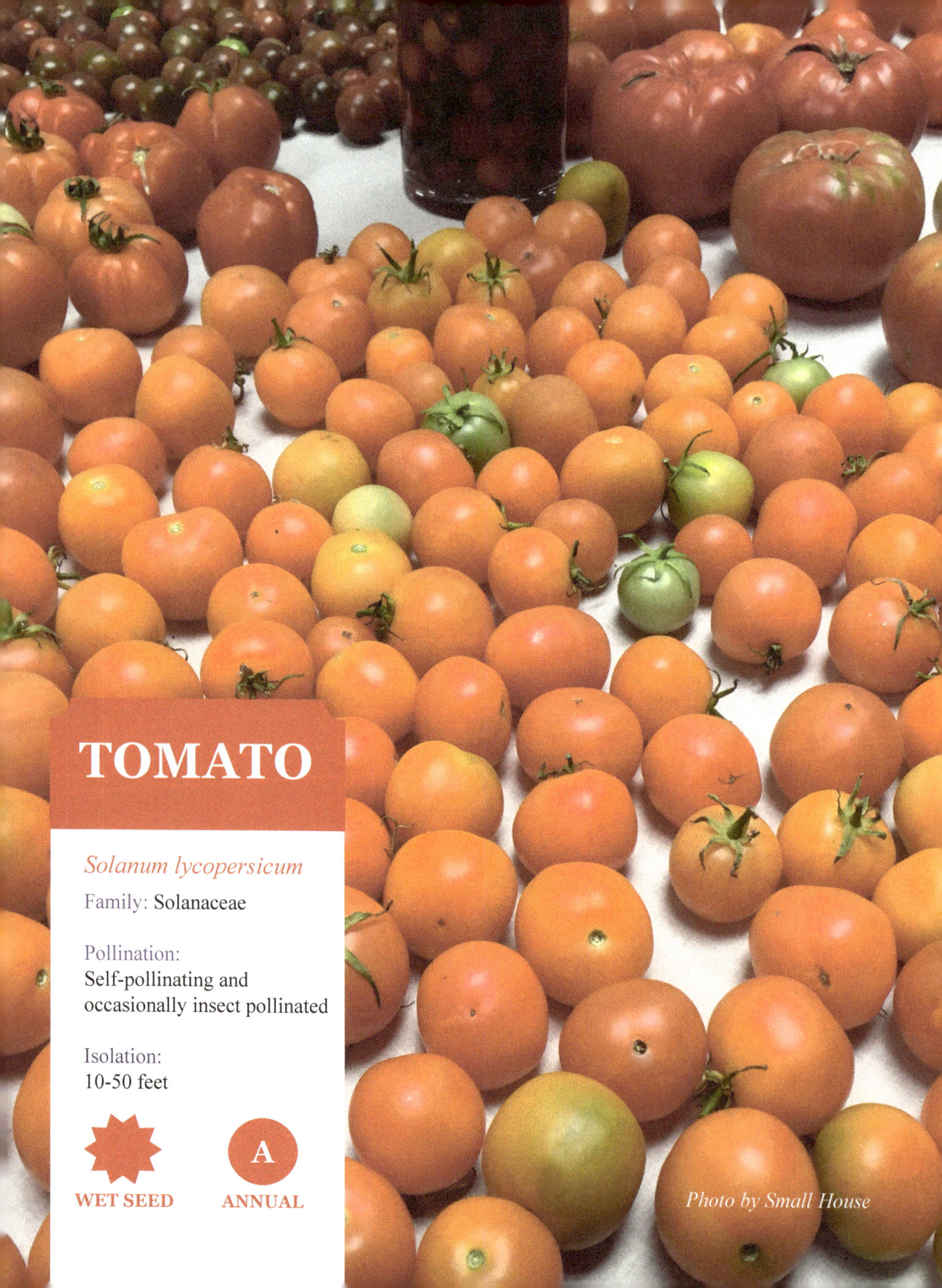

TOMATO

An entire book could be written about the delicious tomato. In fact, there are several volumes on the market specifically dedicated to this amazing and diverse fruit. It may be considered the most popular of all garden crops and every gardener has a favorite variety that they are eager to recommend. Tomatoes come in all shapes, sizes and colors; cherry, beefsteak, oxheart, roma, smooth, lobed, yellow, green, red, striped, bi-color… The plants themselves can either be determinate and grow to a relatively predetermined size with concentrated production time or indeterminate and continue to grow and produce fruits right up until frost brings the season to an end. The leaves of the plants can either have the classic serrated appearance or a smoother margined look known as potato-leafed types. *Solanum lycopersicum* is an excellent example of diversity in the garden!

Tomatoes are, for the most part, self-pollinating although there are some varieties more susceptible to crossing by insects. Discovering what types you may have growing in your garden can be accomplished with just a little bit of observation of the plant's flowers. While most tomato flower's stigmas are hidden away and protected by the anther cone, you will find that some have stigmas that protrude out beyond this protective cone; these are the varieties that are at risk of cross pollination. There is no need to worry though since simply bagging these blossoms to avoid visits from insects will help to ensure that crossing can't occur and your seed will remain pure and true-to-type.

In all tomato varieties, pollination is aided by agitation of the flowers. In nature this is done by insects, bumblebees in particular are perfect for this, or by wind. This can be reproduced in situations without outside inputs by simply shaking your tomato plants gently every few days or by setting up a fan if one is growing in a greenhouse or other similar environment.

When your tomatoes are ripe and ready to harvest, it's also the time to collect the seeds. They can be processed like any other wet seeds but in the case of tomatoes, fermentation is a recommended additional step. You can learn more about the fermentation of seeds on page 176.

SAVING OUR SEEDS: The Practice & Philosophy

SEEDKEEPER'S TALE

LAURA FLACKS-NARROL
MISSOURI, USA
FOUNDER, IVAN TOMATO RESCUE PROJECT

There is a biblical Talmudic quote that states, "Whoever destroys a soul, it is considered as if he destroyed an entire world, and whoever saves a life, it is considered as if he saved an entire world." I believe this to be true for heirloom tomatoes as well.

I started the Ivan Tomato Rescue Project to save a very special heirloom tomato. For many years I grew tomatoes in my suburban back yard and had rather dismal returns. Some of the challenge came from the fact that I live in Columbia, Missouri. Our summer weather was erratic with droughts, monsoon flooding and extreme temperatures, all in one season. Summer after summer I grew tomatoes in an effort to grow my own food and summer after summer my plants died an early, unproductive death.

One year, I complained to a farmer at a local outdoor market. He told me that the Ivan was his family tomato and that I should try it as it could handle the "Missouri Crazy" weather. I took home my first Ivan Tomato. I grew the Ivan and got a wonderfully strong plant that topped out at over eight feet high. It put out lots of good size tomatoes that had a great, old fashioned acid taste. The vines were strong and did not show any signs of the usual diseases. As other plants died in my garden that summer, the Ivan grew and

SAVING OUR SEEDS: The Practice and Philosophy

Photo by Laura Haggarty, Pathfinders Farm

grew and put out more and more tomatoes. We had tomatoes all the way up to the frost.

I saved seeds and continued to raise the Ivan for myself and for some select friends. It was not until several years later that I realized the family, who's tomato this was, had not been at the farmers market in a year or two. Upon inquiry, I found out the father of the family, Jerry, the driving force behind the market booth, had died and the family had not had the emotional wherewithal to continue the nursery business. To make matters worse, they had left all of their seed stock in the hot greenhouse over a couple of summers and all their seeds were no longer viable. This meant that I had the last seeds to the wonderful Ivan Tomato.

There are moments in life where one must make a choice. You can choose to speak up to injustice or stay quiet and say, "It is not my problem". You can choose to achieve great things or sit around complaining. My choice was to either save the Ivan or let the genetics for this wonderful, resilient tomato be gone forever. I decided to grow life, I decided to save the Ivan Tomato and make sure this great heirloom was there for generations to come.

The Ivan is named after Cousin Ivan Koeppel from Memphis, Tennessee. The tomato got its name in an unusual way. The family grew this tomato as their fool proof tomato throughout the family's mid-west origins. They grew the Ivan for canning and eating, saving the best seeds each year. The tomato did not yet have a name; it was just the tomato they all grew.

The story got interesting with this last generation. Jerry, who was the father of the Ivan tomato in the Columbia area, was in the Vietnam War. He was a helicopter pilot and left the war with scars that were not visible on the outside of his body. He had PTSD and eventually had a stroke. The summer after the stroke, Cousin Ivan came up from Memphis for a visit. He had brought some of the seeds from their family tomato. He put them in Jerry's hand and told him to grow life.

SEED KEEPER

Jerry's daughters built him a greenhouse to help him recuperate and grow plants. He started growing wildflowers, then herbs and then fruit and vegetables. This started his plant nursery, Heartland Family Farms. Over the years, fellow veterans would come by his nursery and work with the plants, learning skills and experiencing the healing powers of gardening. Jerry and his family ran this business until he died, enjoying every minute of working with the plants and growing life.

You may think that I am a skilled farmer with plenty of land. This is where you would be wrong. I am a city dweller originally from Toronto, Canada. I did not know how to grow anything when I moved to Missouri in 1996. Now, my husband and I, along with our kids, run a plant nursery that starts out 6000 plants each spring in our kitchen sliding glass door widows. We have a small green house in our back yard as well as many raised beds. We grow food in this suburban homestead and run a farmer's market selling plants, seeds, ointments, fruit and vegetables.

This wonderful heirloom lives on through the Ivan Tomato Rescue Project. We sell seeds Online and we also sell plants at the Columbia Farmers Market and other Mid-Missouri venues, such as Baker Creek's Spring Planting Festival, each spring. While the Ivan's greatness was achieved through its adaptive qualities to the Mid-West, it has also proven itself to grow well in many other microclimates. The Ivan Tomato has been grown in almost every state in the USA, as well as several other countries around the world.

Our company, Victory Gardeners, continues to find family heirloom cultivars to add our rescue project in an effort to save these unique and worthy plants from extinction. The Ivan Tomato is now part of the Slow Food Ark of Taste and has made quite a name for its self. Yet, there are many more cultivars out there also worthy of being saved. If you ever find yourself with the last seeds of a cultivar in your hands; grow life and save an entire world for future generations.

TURNIP

Brassica rapa

Family: Brassicaceae

Pollination:
Insect Pollinated

Isolation:
½ mile

DRY SEED

BIENNIAL

OVER-WINTERING REQUIRED ...SEE PG 184

Photo by Baker Creek Heirloom Seed Company

TURNIP

Like its cousin species, *B. oleracea*, the plants found within the species of *Brassica rapa* are amazingly diverse. Over time these plants have been selected and developed for edible leaves, flower buds and even for their roots! Various plants of this species have been utilized as food by humans for more than 10,000 years.

Although turnips are self-incompatible, they are insect pollinated and will readily cross with other plants of the same species. This includes broccoli raab, mizuna, choy sum, and other Chinese cabbages such as Napa and bok choy. Some of these types are biennial while others are annual, such as bok choy, so some field testing and experimentation may be necessary to help the grower understand these plant's particular life cycles to avoid cross pollination.

Turnips are biennial and will need to be overwintered before they will flower and produce seed. More specific information on the process of vernalization can be found on page 184.

Like many other members of the plant family Brassicaceae, turnips seeds will develop on flower stalks in small pods called siliques. After maturity, these siliques will dry and turn a light brown color indicating that they are ready for harvest. These dried pods are designed to shatter, so the grower must exercise caution when collecting them. Turnips seeds are very easy to thresh due to their proclivity to shattering and winnowing the seeds is also a simple task. More info on threshing and winnowing dry seeds can be found on page 182.

It is also important for the gardener to note that there are varieties of canola that are grown commercially for oilseed production that are of the species *Brasicca rapa*. Cross pollination with these plants is quite likely; awareness and precaution is needed to avoid this. Caging one's plants and introducing pollinators is one option but isolation by distance is much easier to accomplish, as is growing your seed crop on alternating years.

SAVING OUR SEEDS: The Practice & Philosophy

WATERMELON

Nothing says summertime quite like enjoying juicy slices of watermelon at a neighborhood cook-out, while the heat of the day is momentarily quenched by the flavors of this exotic and delicious member of the cucurbit family. Watermelons were originally domesticated in Africa and it is believed that they were cultivated in Egypt around the time of King Tut, 1300 BCE. From there, they made their way to India, then China and eventually into Europe. *Citrullus lanatus* arrived in the Americas around 1550 via European colonists as well as by slaves brought over from Africa.

Like other members of the cucurbit family, watermelon plants are monecious, meaning that each plant has both male and female flowers. While they are self-compatible, watermelons are pollinated by insects and crossing is very likely to occur in gardens where more than one variety is growing. To avoid this, a gardener can choose to grow only one variety of watermelon per season but if this is not possible, or if the garden is in close proximity to others that are growing this species, hand pollination must be used to ensure varietal purity. Instructions on how to hand pollinate flowers can be found on page 172. It's worth noting that the citron melon is of the same species as watermelons and crossing between these two types can occur. In some southern states, citron melons have naturalized and can pose as a cross pollination risk.

When your watermelons are ripe, the seeds are ready to harvest. Knowing when your fruits are finally ripe can pose a challenge, but collecting the seeds is easy. We've all been watermelon seed savers at least once; instead of spitting the seeds onto the ground as you enjoy your summer watermelon, spit them into a container instead, instant seed saving! Be sure to wash the seeds to remove any sugar residue and put them out on screens or paper plates to dry. If you are harvesting watermelon strictly for seed and not to eat, the fruit can be cut lengthwise into quarters and the flesh pressed through hardware cloth into a bucket below. Simply add some water, give it a good stir and let the mixture settle a bit. Next, decant the debris from the top and then pour the seeds, which have settled to the bottom, on to a screen for rinsing.

SAVING OUR SEEDS: The Practice & Philosophy

SEEDKEEPER'S TALE

ROB MCELWEE
LOUISIANA, USA
FARMER & WATERMELON ENTHUSIAST

My earliest memories of gardening are of me helping in my Grandma's flower and vegetable gardens in the early 1980s, when I was around four. She had a large backyard garden that she worked only with hand tools. I would help her plant seeds and transplants and pick the vegetables. The first watermelons I grew were a single hill of 'Crimson Sweet' which produced only two melons that year but unfortunately, one was picked green and I dropped the other on the patio and it busted open. I still ate it though!

My love of watermelons truly began with my uncle, Baron Clinton. Every year he would plant over a hundred hills of watermelons along with muskmelons and cowpeas. He would pick his melons and pile them in the shade of huge old plums and mimosa trees. When my grandma and I would visit, I would run over to the piles of melons, amazed by the diversity; large, medium and small ones, round, blocky and long melons, striped, dark green, gray and almost white on the outside, yellow, orange, pink and red flesh on the inside. I would stand and stare a moment and then dive in among them, rolling around on top of them and laying on them, feeling the coolness that they radiated on the hot summer day. My uncle would load my grandma's small truck with a dozen or so of the ones I picked out and when we got

SAVING OUR SEEDS: The Practice and Philosophy

Photo by Small House

home we would line them up in the breezeway and I would play with them and draw on them with markers and every other day we would cut the one I picked out for us to eat.

As I got older, I would ask my uncle the names of the watermelons he grew. Tom Watson, Kleckley Sweet, Graystone, Charleston Gray, Black Diamond red and yellow, Jubilee, Tendersweet, Desert King, Irish Gray, and Dixie Queen are the ones I remember but I am sure he grew more. This went on for years but I, as young as I was, noticed that over the years my uncle grew less and less varieties. When I was around ten I asked him why. He told me the feed store where he bought most of his seeds had quit selling the old ones that he liked to grow and several of the catalogs he ordered from had gone out of business or stopped selling them. He would tell me tales of when he was young, in the 1910s, when farmers would plant fields of watermelons that are no longer grown; the size, rind colors and shapes were amazing. He spoke of how wagons would take them to train depots and that they would be loaded onto trains and shipped north. He would tell me of stealing watermelons as a kid that were so big that it would take him and a friend to carry them out of the field. He told me that stealing a watermelon was a rite of passage for a kid back then. Farmers expected it and if you got caught, you failed and you got a whipping, but not too bad of one.

My grandma began to plant bigger gardens and my uncle smaller ones as he got older. She started planting lots of watermelons. I was disappointed that at the local feed stores and seed racks I could only find Jubilee, Charleston Gray, Black Diamond, Crimson Sweet, Tendersweet and the super common Sugar Baby. I saw for the last time, in 1989, Kleckley Sweet on a seed rack. In 1990, I got my first Hastings seed catalog and I ordered Graystone and Stone Mountain from it and grew them out that year. I had a great year but at the time I knew nothing of hand pollination so I ended up with a bunch of crossed up melons. In 1992, I got my Hastings catalog and intended to order them again but they had dropped Graystone. I was shocked. I was very upset at the thought of not being able to grow it again.

At the same time, I had learned of Seed Savers Exchange and joined. I got their yearbook and the Garden Seed Inventory Second Edition. I was enthralled by the number of watermelons still offered and saddened by the list of ones no longer being sold, many of my uncle's favorites. I had also learned about hand pollination from a vegetable breeding book that was given to me. I made it my mission to acquire every one of the watermelons still being offered in those catalogs. I started mowing grass and doing chores around the neighborhood and selling bushels of cowpeas to raise money. I bought stacks of postcards and began requesting the free catalogs and mailing money orders for the ones that cost money and sending SASEs (self-addressed, stamped envelopes) for price lists. And then the ordering began. Packages started arriving in January. I decided to plant them on a five year rotation because I only had a 150' x 150' area to plant, so the melons I was not going to plant that year were stored in glass jars in the freezer.

When the Garden Seed Inventory Third Edition arrived, I ordered all the new watermelons in it. I didn't request many from the SSE yearbooks because I had found many were crossed badly. 99% of the people listing there had no idea about keeping lines pure through isolation or hand pollination. Keeping things pure is difficult for most crops, with a few exceptions. Tomatoes and beans don't cross at the same rates as squash and melons.

In 1992, at age 15, I got my driver's license and that opened up a whole new world. I began to explore the areas around where I lived and the old areas of Natchez and Bermuda, Louisiana. I found a treasure trove of old vegetables; cowpeas, muskmelons, a rare red eggplant, tomatoes, sweet potatoes, okra and, of course, watermelons. The watermelon diversity was amazing. For example, a very large, oblong dark green watermelon with bright yellow flesh and very black seeds, a pale green version of Moon & Stars that had been in that family for 100+ years, a very small, very dark green melon with bright orange flesh, one that looked like Charleston Gray but with white seeds and many more. These seeds were being maintained by very elderly people who said that when they could no longer grow them, the seed would most likely be lost because their children and grandchildren were not interested in gardening. These elderly gardeners knew all about crossing and only planted one variety of watermelon a year. I was amazed at their knowledge and took notebooks full of notes and recorded conversations. They loved to talk and were glad that I was interested in growing their old varieties.

By 1999, my collection had grown to be huge; 258 watermelons, 86 muskmelons, 112 pepo and moschata squashes, 56 cowpeas and several hundred other vegetables. I had my system down and my fields were well worked. I had survived droughts, deer, raccoons, coyotes and people raiding my garden. I had bought three freezers to store my seeds in and kept my notes and tapes in totes in the family storage shed. Then it was all destroyed.

In early March of 2002, I was hooking my trailer up to go get a load of composted cotton waste when I noticed my neighbor, stumbling around his field, obviously drunk. I paid him no mind because that was nothing unusual with him. I should have noticed that he was staring at a huge pile of pine tops from when he had his pine trees cut the previous year. I was in a hurry because the loader of the compost was waiting for me. As I drove down the driveway, I had a nagging feeling that bothered me. As I was heading back home I noticed a black column of smoke in the distance and something told me to speed up, my truck protesting at having to pull the heavy load faster. As I turned onto my street I could see fire trucks and cops turning into my driveway. I floored it and turned in right behind them and jumped out of my truck. My storage shed was a raging inferno. My neighbor had set fire to his piles of brush without thought of fire breaks and the fire had spread everywhere in a matter of minutes while he stumbled back to his house to pass out.

I watched as the fire consumed ten years of hard work. I felt ill. The fire burned so hot that the metal siding melted and pooled on the ground. All my notes, rare old Organic Gardening magazines, pictures, my taped conversations with those old gardeners and, of course, all my seed were destroyed. The fire thawed the deep freezers and they got so hot that the jars shattered. I was devastated. The neighbor was ticketed and his insurance covered the buildings but nothing could replace what I had lost. I gave up. I just didn't have the heart or will to build my collection again. Besides, numerous varieties of watermelons had been dropped from the commercial catalogs and were no longer available. All of the elderly people I had talked to and gotten seeds from were either dead or in nursing homes. Also, I had a full time job and lacked the time to do it again. I still raised vegetables but made no effort to save seeds anymore. Instead, I began to concentrate on growing old roses and flowers for the next sixteen years.

Then, in 2018, I was looking through my old Garden Seed Inventory Third Edition and, on a whim, began to see how many varieties listed in it were still available. What I found stunned me. Only around 20% of those listed in 1992 were still available! I was shocked. So I pulled my I, II, IV, V and VI editions, did the same and added up my results. The outcome was stunning. Of all the varieties listed in those six books, only 40% of the varieties were still available. The results were even worse for the newer, open pollinated varieties. Additionally, I learned that most seed companies listed in the inventories were long gone and the unique varieties that they had offered were extinct. Seed companies that specialized in watermelons, such as Roswell Seeds and Burrells, dropped the rare ones they offered and gone mainstream. Another great loss was Wilhite, for example. In the 1980s and early 1990s it was one of the leading growers of heirloom and new, open pollinated and hybrid watermelon seed. They offered many rare watermelons such as Black Diamond Yellow Flesh and White Seeded Watson that were not offered anywhere else at the time. Now, they only offer a handful of common, open pollinated varieties and nothing rare. I saw that some watermelons went from twenty sources for seed, to one or two and some even rarer varieties go from five or six sources to only one! I began searching for the old varieties and found very few left on the Internet but found lots of new companies which were offering a limited selection of open pollinated watermelon varieties.

Something stirred within me and I knew I had to take action. I began ordering every open pollinated watermelon I could find because seed companies seem to come and go really fast. I literally ordered Merrimac Sweetheart, a watermelon that hasn't been offered in 70+ years and that I did not have in my old collection, from a little Georgia seed company and after I got my seeds (which were among the best packaged) I tried to place another order for some rare okra and their website had been shut down. If you hesitate you may never have another chance, so if you see something you want, you'd better get it. I ended up with 159 open pollinated varieties by the end of that March. I had planned to grow out 50 varieties this year but due to excessive rain and a very late start, I only planted ten of my rarest ones. I hope that they grow because four of those varieties are extremely rare. One is said to be the long lost variety Ravenscroft but I won't know until I see it.

I have been doing lots of research on watermelon varieties from the past and I hope to write a book on the history of the watermelon in America and list descriptions of all the varieties I can from 1800 to today.

Photo by Laura Haggarty, Pathfinders Farm

WHEAT

Triticum spp.

Family: Poaceae

Pollination:
Self-pollinating, wind pollinated occasionally

Isolation:
10-20 feet

DRY SEED

A ANNUAL

Photo by *Great Lakes Staple Seeds*

WHEAT

While it's true that wheat may not be as colorful as tomatoes or peppers, or quite as easy to incorporate into the kitchen routine as kale or cantaloupe, wheat is the basis of the food pyramid and the movement towards a local food system can only be as successful as our ability to cultivate and harvest our basic staple crops. While wheat may require a bit more space than other more commonly grown garden plants, it is easy to grow, simple to harvest and a pleasure to enjoy.

There are a few species of wheat in commercial production today with the primary market being held by *Triticum aestivum* which includes hard white, soft white, hard red and soft red types. Additionally, there is *T. turgidium* which is also called durum wheat, most well-known for its use as semolina flour for pasta. Einkorn and Emmer wheats are also each of their own distinct species, yet all four of these are harvested and processed for seed in the same manner.

Wheat flowers are perfect and self-pollinating but on rare occasion can also be pollinated via wind. Crossing between wheat varieties is quite rare and little precaution or isolation is needed. Seeds of wheat are harvested in the same manner as they would be for consumption, simply allow the plants to mature and dry in the field until the kernels are hard and unable to be dented with a thumbnail. When mature, a small plot of wheat can easily be harvested by hand using a scythe and placed somewhere protected from the elements to continue drying for a few more days.

When ready, the seeds can easily be threshed from the plants by flailing, treading or otherwise beating them loose. More detailed information on threshing and winnowing dry seeds can be found on page 182.

I highly recommend that every gardener, regardless of the size of their plot, should try growing wheat at home at least once. While you may not have the space to provide for your family's annual bread needs, growing and harvesting even a small amount of wheat is a satisfying endeavor that everyone should experience.

SAVING OUR SEEDS: The Practice & Philosophy

Modern Trade Routes
By Bevin Cohen

As kids, we all learned in school about the ancient trade routes used to move spices, tea and other exotic commodities around the world. The most well-known was the Silk Road, of course, but it was the maritime spice routes that brought explorers around the globe and even played a role in Columbus' "discovery" of America. Now you and I both know that someone can't discover land that someone else already lives on, but that's another topic for a different book.

When learning about these trade routes, we also learned that they served a purpose beyond just trade between countries. These roads also facilitated cultural exchange: religion, ideas and knowledge. But unfortunately, we also learned about these trade routes in the past tense.

As a seed keeper, I serve multiple roles such as gardener, historian, educator and story teller but through this work I also often find myself on the road traveling extensively to share these precious heirlooms with my fellow seed savers wherever they may be. We gather at seed swaps, farm conferences, libraries, coffee shops; basically any place that offers enough room for us to spread out our treasures on a blanket for display. In a way, we are quite similar to these ancient tradesmen that traveled across the land offering their exotic herbs and spices, their teas and jewelry, to curious and interested onlookers. We too are eager to share our unique and valuable items that have sometimes traveled a great distance to be admired and appreciated by gardeners from a far-away land.

A couple of years back, I acquired seeds of a beautiful bean from a friend named Debbie Groat that lived not too far away from me in Michigan. These were beautiful seeds for a variety of pole bean known simply as Grape. As the name implies the bean is spherical, almost perfectly round, and a deep dark red color. These seeds were absolutely gorgeous and Debbie was actually using it to make jewelry! When I first saw them, I knew that I just had to have them. How could I resist their siren song? Thankfully Debbie is a generous friend and she happily sent me home with a sample of these incredible beans.

I grew the beans that summer and they did quite well for me. They were vigorous and healthy plants that grew vines six to seven feet long, taller than the teepee I had built for them in our front garden. The pods themselves were rather flat, each containing six to eight round little seeds that quickly filled out their shells. The brown and yellow mature pods were ready to harvest by the end of September and I was blessed with a large harvest of gorgeous, round, nearly purple seeds. I'm always amazed at how prolific plants can be, from just a small handful of seeds came well over a pound of beans for me to enjoy and share with my friends.

The following spring, I was back on the road to visit friends and swap seeds at one of my favorite events, the Appalachian Seed Swap. I was excited to show off my recent harvest of Grape beans as I was sure that they'd be a stunner. They did not disappoint. Everyone was in awe of these incredible seeds and I soon started to wish that I would have brought more of them with me because they were going fast! Everybody was eager to get a sample of the seeds and I was more than happy to oblige until one older lady stopped at my table, eyed my beans with a most inquisitive stare and then asked in an almost accusatory tone, "Where did you get these seeds?"

After a brief, and maybe nervous, hesitation I told my visitor the story of my friend Debbie and the beautiful seed jewelry that she was known for. I asked her to reach out her hand and into it I poured a small pile of these burgundy-merlot colored pearls. I wasn't exactly sure what to expect in response but I never would have guessed the incredible tale I was about to hear. As it turns out, the lady at my table recognized this unique bean and she remembered her grandmother growing this same variety when she was young girl. Grape is what is known as a fall bean, meaning that the seeds are ready to be harvested and used as a dry bean late in the growing season. As she gazed reminiscently at the seeds in her hand, she stirred them around with her finger and recalled that her grandmother used to make the most delicious baked beans dish for Sunday dinner using the Grape beans that grew in her home garden.

I inquired as to what had happened with her grandmother's beans and if they grew them anymore. Her voice turned sad as she told me about how her parents had moved away when she was still a young girl and she wasn't sure what had happened to the family seeds. Her folks had maintained a garden for a short time after they had settled into their new home, but when she and her brothers had grown up, they moved on and didn't keep a garden anymore. Unfortunately this is a story that I hear all too often when I'm trying to track down the history of a family heirloom variety. Too many times the trail runs cold and we can only follow the story back so far. I asked the lady where her family had moved to when they left the Appalachian Mountains, "Up to Michigan, of course, we had to go where the jobs were. My daddy had gotten a job at the automobile plant."

Was it possible that the beautiful red seed that I had acquired from my friend in Michigan was the very same variety that this lady's family had brought with them from Kentucky so many years before? And was it then possible that this seed had followed this modern trade route back down to find its way home here in the Appalachian Mountains? One of the lessons I've learned after years of working with seeds is that there's no such thing as coincidence. Just as the ancient trade routes linked people, countries and cultures together, the modern trade routes of the seed keeper are continuing this tradition of connecting people, helping us to build the bridges of commonality that are so desperately needed in today's world. Through seeds, we can realize that we are all one people. Through seeds, we can share and express our love of the natural world. Through seeds, we see that we are all connected. These modern trade routes are the webs through which we are all entwined.

The Practice

Flower Structure and Pollination

As seed savers, it's important that we devote the necessary time to developing proper relationships with our plants. Even the simple act of observation allows us to learn so much from the crops we tend and the fruits that we harvest. Our studies must begin at the same place where the life cycle of our precious seeds begins; the flower.

This beautiful pepper blossom is an excellent example of a perfect flower.

Flowers can best be described as the reproductive organs of a plant. This is where pollination, fertilization and fruit and seed development happens. The parts of a flower are commonly divided into male and female parts. The male part of the flower is the stamen which consists of the anther and the filament. The female portion of the flower is known as the pistil and is made of the stigma, style and ovary. It is the ovary which will eventually grow to become the fruit, after proper fertilization occurs.

A number of our garden plants have perfect flowers, which mean that they consist of both male and female parts. Quite often these plants are self-pollinating, although at times this is not the case and the flowers are self-incompatible. Many members of the Brassicaceae family (broccoli, cabbage, collards etc) are great examples of this. Self-incompatible

A conjoined tomato flower, also called a super bloom, is more susceptible to cross pollination.

SAVING OUR SEEDS: The Practice & Philosophy

Flower Structure and Pollination

flowers require pollen from other flowering plants of the same species in order to develop fruit and, eventually, seeds.

Conversely, a fair amount of our popular garden crops are monoecious, meaning that each plant will have separate male (staminate) and female (pistillate) flowers. Squash, melons and cucumbers are perfect examples on monoecious flowers. Corn is another monoecious plant; the tassels that form at the top of the stalk are the staminate flowers, while the silks growing further down the stalk near a leaf node are part of the pistillate flowers. Other times our plants will be dioecious, having plants each bearing either staminate or pistillate flowers. Spinach is a fine example of a dioecious plant species.

In addition to this, although some of our flowers are self-pollinating, many others require outside assistance from either wind or insects to ensure that pollination and fertilization will occur. It's important that we understand how our flowers are being pollinated so we, as seed savers, know if and when we need to get involved. Observe your plants, study their flowers and watch the magic of their life cycle unfold.

A female squash flower (left) can be easily identified by the small fruit located beneath the petals as opposed to a male flower (right).

Many happy pollinators visit this cucumber flower.

Hand Pollination

For many gardeners hoping to save seeds from their crops, it can be challenging to achieve the recommended isolation distances between varieties of the same species to avoid cross pollination and ensure pure, true-to-type seed for use the following season. One technique that can be utilized successfully with a bit of practice is the art of hand pollination.

Some of the most common crops that are hand pollinated by home gardeners are members of the Cucurbitaceae family which includes summer and winter squashes, cucumbers, melons and watermelons. While there are certainly some differences from one species to the next, such as the size of flowers, each of these plant's fruits are pollinated in the same manner and therefore the same technique can be applied to each.

These plants are monoecious, meaning they have separate male (staminate) and female (pistillate) flowers. These are easy to identify as the pistillate flower has a visible ovary just below the petals that resembles a miniature version of that plant's fruit, while the staminate flower does not.

Hand pollination of Curcurbitaceae flowers is a two day chore. On the first day, the prospective seed saver will head out into the garden in the evening to identify which flowers will likely be opening the next morning. In many cases, the visual clues that a flower will be opening are easy to find; the fused tips of the petals will be beginning to split and the yellow coloration of the flowers will be more visible.

Once the flowers have been chosen they must be secured shut to avoid having them open before the gardener returns in the morning. This can be accomplished using clothespins, masking tape or even twine. Be sure they are well secured and the petals have not been ripped. For the female flower, a clothespin is an ideal tool for this step since we do not wish to damage the petals when opening them the next day. Be sure to mark where these flowers

Hand Pollination

are located in the garden to expedite relocating them in the morning; small utility flags are great for this.

The following morning, locate the male flowers that you had previously taped shut and remove them from the plant. Take them with you as you relocate a female flower. Expose the anthers of the staminate flowers by removing and discarding the petals and then gently open the pistillate flower to expose the stigma. Using the male flower like a paintbrush, gently brush the anthers onto the entire surface of the stigma. Use two or three staminate flowers for each pistillate flower if possible. Once this is accomplished, reclose and secure the petals of the female flower to avoid contamination from contact with other insects or stray pollen. Be sure to mark the stem directly below this flower to ensure proper identification of the hand pollinated fruit later in the season. Yarn or ribbon will work well for this.

Bagging Blossoms

Photo by Small House

Bagging blossoms is a simple and effective method to isolate self-pollinating crops to insure varietal purity. Simply place small mesh bags over the flowers before they bloom to avoid interactions with insects. Being self-pollinated, your fruits will still form inside of the bag. It is safe to remove the bag after fruits have set.

Hand Pollination

While squash, melons and cucumbers may be some of the more common garden crops that are hand pollinated by prospective seed savers, they are certainly not the only ones. Corn, *Zea mays*, is a popular summer favorite that can easily (with some practice) be hand pollinated to help ensure varietal purity.

Corn plants are also monoecious; the tassels on the top being the pollen shedding male inflorescence and the silks that form near the leaf nodes are part of the female flower. Just as with our cucurbits, the first step is cover the staminate flower to avoid pollen shed and protecting the new female flower from uncontrolled pollination. This can be accomplished by placing a small paper bag over the emerging corn tassels to collect the plant's pollen as it is dropped. The seed saver will also want to place a small bag over the young silks to avoid unwanted, accidental pollination from another variety. While it's possible, and recommended, to purchase waxed pollination bags made especially for this purpose, one can also simply use small brown paper bags if they are unable to make this purchase for whatever reason.

Photo by Laura Haggarty, Pathfinders Farm

The silks that form near the leaf nodes of the corn plant are part of the female flower.

Hand Pollination

In the morning, after the dew has evaporated and the pollination bags are dry, the gardener can gather the pollen bags from the corn tassels by bending the plant downward and removing the bag while remembering to shake the plant gently to remove any extra pollen. Once all of the corn pollen is collected it can be mixed together, to ensure good diversity in your crop, and then used to pollinate the silks that had also been bagged the day before. Simply remove the protective bag, evenly sprinkle the pollen upon the exposed silks and then re-bag them right away to avoid contamination from unwanted pollen sources. These bags should be left on the plants until the silks have dried down and turned brown and can be left in place until harvest to help identify the hand pollinated ears.

The tassels that grow at the top of the corn plant are the pollen shedding male inflorescence.

Hand Pollinating Squash
Step by Step

Photos by Small House

1. In the evening, locate male flowers and tape them shut.

2. In the evening, locate female flower and gently clip closed.

3. The following morning, remove male flowers from plant and remove petals to expose stamen.

4. Carefully open female flower and brush the stigma with male flowers.

5. Gently reclose and tape female flower, being cautious not to damage petals.

6. Clearly mark your pollinated flower with string to help you find it later in the season.

Processing Wet Seeds

When seeds that are ready to be harvested are contained within a fleshy fruit, they are typically referred to as wet seeds. Common examples would include melons, squash, cucumbers, tomatoes, peppers and eggplants, as well as many others. Gathering and processing these seeds requires a few simple steps and in the case of tomatoes, cucumbers and a few others, additional time for fermentation is needed.

It is important to wait to harvest your seeds until the fruits themselves are physiologically mature. Many times, we pick our fruits for consumption or market when they are still technically immature; waiting for maturity increases the chances of harvesting healthy, viable seeds.

Once the grower has waited until the seeds of their particular species are mature, harvesting is simply a matter of cutting open the fruit and removing the seeds. Next, the seed saver will want to clean the seeds to remove any flesh from the fruit as well as any residual sugars to avoid potential mold issues that may arise. This can be accomplished on a small scale with a colander under the kitchen faucet or with seed cleaning screens and a hose or sprayer for larger harvests. Once washed, the seeds should be put out to dry either on screens or even on paper plates, which are easy to label with the variety's name. The gardener can then use a fan on a low setting to expedite the drying process.

Small fruits such as ground cherries and potato berries can quickly be run through a food processor to remove their seeds. Add a bit of water with the fruits and whir them a bit using a plastic dough blade to reduce the risk of damaging the small seeds. These can then easily be decanted to remove the debris that floats to the top, while keeping the mature seeds that sink to the bottom of the water.

Photos by Sow True Seed

Fermenting Seeds

As mentioned earlier, some seeds such as cucumbers and tomatoes can benefit from the additional task of fermentation. While this may sound challenging or difficult, it's actually quite easy. When the seeds for your particular species are mature, collect them from the fruits and place them in a jar or small bucket and add a small amount of water. Next, cover with cheesecloth to keep the fruit flies out while still letting your concoction breathe. Allow the containers to sit for three or four days until a white mold begins to form on the top. Once this happens, add a small amount of water and give the jar a quick stir; after a few seconds the contents of the container will separate with the healthy, mature seed sinking to the bottom and everything else then floating to the top. Simply decant and discard the undesired portion, strain and rinse your freshly cleaned seeds, then place them out on a screen or paper plates to dry.

The fermentation of these seeds, while not 100% necessary, is highly recommended. This process helps to remove growth inhibitors that are naturally present in the gel-like coating on the seeds, while also preventing disease from being carried over from the previous season. Additionally, immature seeds tend to float so they are also removed during this processing of wet seeds.

Fermenting Tomato Seeds
Step by Step

Photos by Sow True Seed

1. Gather tomatoes. Label containers to avoid confusion between varieties.

2. Cut tomatoes in half horizontally to reveal the seed cavities.

3. Squeeze or scrape tomato seeds into a container.

4. Add water and stir. Leave to ferment 3-4 days. Seeds will sink to the bottom.

5. Pour off the fermented debris at the top and rinse the remaining seeds.

6. Lay seeds on a paper plate, screen or coffee filter to dry before storage.

Threshing and Winnowing Dry Seed

After dry seeds have been gathered from the garden, they typically need to be separated from the plants and cleaned of debris. These techniques are respectively known as threshing and winnowing.

Removing our precious seeds from mature and properly dried plants through the process of threshing is a very simple and straightforward task. While larger farm operations utilize threshing machines, the small scale seed producer can make use of a number of common household items to get the job done.

Small harvests can be easily threshed by hand, especially large seeded crops such as beans and peas. One can also simply take a handful of dried seed stalks and firmly bang them against the inside edge of a bucket; this will bust apart the pods and the seeds will fall into the bucket. Another option is to lay the harvest out on a tarp and bust the seeds free using a flail or even by walking and stomping upon the plants. A few years ago, I attended an event where we processed wild rice by donning leather moccasins and dancing upon the harvest to release the rice seeds from their shells.

Once the seeds have been broken free from their plants, they must be winnowed in order to remove any additional plant material. This can be accomplished in a number of ways, depending upon the resources available to the grower. By far the simplest way to winnow seeds is by using the wind. At the rice harvest event we all gathered around the seed covered tarp after

SAVING OUR SEEDS: The Practice & Philosophy

Threshing and Winnowing Dry Seed

the dancing was over, took the edge of the tarp in our hands and proceeded to methodically and rhythmically launch the seeds up into the air, only to watch them land safely upon the tarp below. With each toss, the wind carried away a bit of the debris and the seeds, being heavier, fell back down to be collected when we were finished.

This basic technique can be replicated using a fan to replace the wind. The seeds are poured from a height in front of the fan, the unwanted shells, stems and leaves are blown away and the seeds land safely in a container below. On a smaller scale, one can also winnow by using their breath and a small bowl to toss the seeds in. This works particularly well for very small seeds.

Photos by Sow True Seed

Alternatively, one can clean small seeds using a series of screens. The size of the screen mesh is chosen so that the seed is small enough to fall through, but the plant debris is too large to fit. This may take a few rounds of sifting to get the job done or a small amount of winnowing may still be needed to fully clean the seed harvest.

Overwintering Biennials

A few of the garden crops that one might choose to also grow for seed production are what are known as biennials. The word biennial simply means "two-year", as opposed to annual plants that complete their life cycle within the first year of planting. These particular biennial species require a period of overwintering, or vernalization, in order to trigger the plant to begin its reproductive cycle, flower and eventually produce seeds. Typically, a biennial plant will need to be exposed to temperatures between 40-50°F (4-10°C) for a period of weeks to induce flowering.

In some areas of the world, where winter temperatures are mild, this can be accomplished by simply leaving the plants in the garden to be exposed to the cold weather. The grower will want to plant these crops a bit later than they would regularly plant them for a food harvest so that the specimens are still young and strong when the winter weather arrives. A good layer of straw or leaves for mulch will help keep these young plants alive; row cover is also a viable option.

In areas where the winters can reach well below freezing, the prospective seed saver will want to dig up their biennial crops to be stored until the following spring. Again, it's best to plan to overwinter plants that are still young and vibrant when they are dug up; this increases the likelihood that they will survive storage. Be sure to trim away any excess leaf material as this will decay and will likely cause your vegetables to rot while being stored.

Plants can be stored over the winter in a root cellar, or packed in moist sawdust or sand. Damp potting soil can also be used. Replant your overwintered crops in the spring, as soon as the soil can be worked. Be sure to give your plants extra spacing to accommodate for flowering and seed

Overwintering Biennials

production, some crops can take up significantly more area in this stage of their life then when grown simply for eating.

Areas that don't reach temperatures of 40-50°F (4-10°C) at night can choose to vernalize their biennial crops by placing them in the refrigerator to simulate the winter weather needed to trigger flowering.

Photo by Baker Creek Heirloom Seed Company

Labeling and Storage

After all the hard work the gardener puts into growing, isolating, harvesting and processing their seed crops, this final step is crucial to ensuring that these precious seeds will retain their vitality and will remain able to be grown and shared for many seasons.

The importance of properly labeling your seeds cannot be understated; without proper documentation and organization, the gardener will quickly find themselves growing a field of mystery crops. For example, the many varieties of tomato may have a multitude of identifiable outward appearances but the seeds themselves all look nearly identical. Proper labeling throughout the entire season is critical. Not only do your crops need to be labeled in the field, this identification must continue from harvest through seed saving.

Some tricks and tips that may help the eager seed saver keep their varieties straight include writing the varietal name directly on the harvested fruits with a permanent marker, labeling the rafters in the shed where seed pods have been hung to finish drying and writing on the paper plates or screens being used for drying wet seed harvests. Some of our large seed crops are dried on screen doors placed horizontally on sawhorses in the pole barn; just a small square of chalkboard paint on the wooden edge of the door allows me to label and relabel my harvests throughout the entire season.

Once your seeds are completely dried, they are ready to be put away until next season. While storage methods vary depending upon one's need, space and available resources, the three most important details are to keep seeds cool, dark and dry. This could be as simple as putting your seeds in coin envelopes, then into a shoe box and storing them in a closet in your basement. Many of our seeds are kept in mason jars in the same part of the house where we keep our canned foods, potatoes and winter storage squash.

Labeling and Storage

• •

Keeping your seeds in airtight containers in the refrigerator is another great option. Seeds stored in this cool environment, where temperatures rarely fluctuate, can remain viable for a number of years. The ultimate long term storage solution, for those that are able, is to place one's properly labeled seeds in airtight containers in a freezer. Seeds stored below freezing temperatures will keep for many years.

Photo by Small House

Suggested Reading

There are hundreds, if not thousands, of wonderful gardening books on the market, but I certainly couldn't include them all here. These are the books that come down from the shelves most often in my home, all of which are inspiring and entertaining:

100 Heirloom Tomatoes for the American Gardener by Carolyn Male
Beautiful Corn by Anthony Boutard
Breed Your Own Vegetable Varieties by Carol Deppe
Buffalo Bird Woman's Garden by Gilbert Wilson
Collards by Edward Davis & John Morgan
Eat Here by Brian Halweil
Epic Tomatoes by Craig LeHoullier
From Our Seeds & Their Keepers by Bevin Cohen
Garlic is Life by Chester Aaron
Heirloom Vegetable Gardening by William Woys Weaver
Homegrown Whole Grains by Sara Pitzer
How Plants Get Their Names by L.H. Bailey
Kentucky Heirloom Seeds by Bill Best
Melons for the Passionate Grower by Amy Goldman
People With Dirty Hands by Robin Chotzinoff
Peppers of the Americas by Maricel Presilla
Plant Breeding for the Home Gardener by Joseph Tychonievich
Saving Seeds, Preserving Taste by Bill Best
Seed Libraries by Cindy Conner
Seedtime by Scott Chaskey
The Devil's Dinner by Stuart Walton
The Heirloom Tomato by Amy Goldman
The Story of Corn by Betty Fussell
The Triumph of Seeds by Thor Hanson
The Whole Okra by Chris Smith
Uncommon Fruits Worthy of Attention by Lee Reich

About the Author

Bevin (Ben) Cohen is a writer, poet, herbalist, gardener, seed saver and wanderer. He lives and works at Small House Farm with his wife, Heather, and two sons, Elijah and Anakin, on their family homestead in Sanford, Michigan.

The Cohen family grows seed crops, herbs and flowers and keep a small flock of laying hens. Bevin is also the founder of MI Seed Library, a collaborative seed sharing initiative that has worked closely with a number of communities to help establish seed library programs across his home state and beyond.

He spends his time outside of his gardens offering workshops and lectures across the country on the benefits of living closer to the land through seeds, herbs and locally grown food. Bevin finds himself at home wherever his hands are in the soil, sowing the seeds of yesterday to reap the bountiful harvest of tomorrow.

Other books by Bevin

Throughout his travels across the country, author Bevin Cohen has collected many interesting and heartwarming stories about heirloom and heritage seeds as well as the people that keep them. This book gives voice to these sacred tales and is told in the words of the seed keepers themselves; a unique blend of history and philosophy. *From Our Seeds & Their Keepers* is a one of a kind collection of stories that is sure to inspire every gardener to take part in the ancient ritual of seed saving.

Available at www.smallhousefarm.com